T0254647

THE FRONTIERS COLLECTION

Series editors

Avshalom C. Elitzur
Iyar The Israel Institute for Advanced Research, Rehovot, Israel
e-mail: avshalom.elitzur@weizmann.ac.il

Zeeya Merali
Foundational Questions Institute, Decatur, GA 30031, USA
e-mail: merali@fqxi.org

T. Padmanabhan
Inter University Centre for Astronomy and Astrophysics (IUCAA), Pune, India
e-mail: paddy@iucaa.ernet.in

Maximilian Schlosshauer
Department of Physics, University of Portland, Portland, OR 97203, USA
e-mail: schlossh@up.edu

Mark P. Silverman
Department of Physics, Trinity College, Hartford, CT 06106, USA
e-mail: mark.silverman@trincoll.edu

Jack A. Tuszynski
Department of Physics, University of Alberta, Edmonton, AB T6G 1Z2, Canada
e-mail: jtus@phys.ualberta.ca

Rüdiger Vaas
Bild der wissenschaft, Redaktion Astronomie, Physik, 70771
Leinfelden-Echterdingen, Germany
e-mail: ruediger.vaas@t-online.de

THE FRONTIERS COLLECTION

Series Editors
A.C. Elitzur Z. Merali T. Padmanabhan M. Schlosshauer
M.P. Silverman J.A. Tuszynski R. Vaas

The books in this collection are devoted to challenging and open problems at the forefront of modern science, including related philosophical debates. In contrast to typical research monographs, however, they strive to present their topics in a manner accessible also to scientifically literate non-specialists wishing to gain insight into the deeper implications and fascinating questions involved. Taken as a whole, the series reflects the need for a fundamental and interdisciplinary approach to modern science. Furthermore, it is intended to encourage active scientists in all areas to ponder over important and perhaps controversial issues beyond their own speciality. Extending from quantum physics and relativity to entropy, consciousness and complex systems—the Frontiers Collection will inspire readers to push back the frontiers of their own knowledge.

More information about this series at http://www.springer.com/series/5342

For a full list of published titles, please see back of book or springer.com/series/5342

Bernd-Olaf Küppers

The Computability
of the World

How Far Can Science Take Us?

 Springer

Bernd-Olaf Küppers
Munich
Germany

Translated by
Paul Woolley
Berlin
Germany

ISSN 1612-3018 ISSN 2197-6619 (electronic)
THE FRONTIERS COLLECTION
ISBN 978-3-319-88421-9 ISBN 978-3-319-67369-1 (eBook)
https://doi.org/10.1007/978-3-319-67369-1

The original version of this book was published in German under the title „Die Berechenbarkeit der Welt: Grenzfragen der exakten Wissenschaften" in 2012 by S. Hirzel Verlag (Stuttgart).

© Springer International Publishing AG 2018
Softcover reprint of the hardcover 1st edition 2017
This work is subject to copyright. All rights are reserved by the Publisher, whether the whole or part of the material is concerned, specifically the rights of translation, reprinting, reuse of illustrations, recitation, broadcasting, reproduction on microfilms or in any other physical way, and transmission or information storage and retrieval, electronic adaptation, computer software, or by similar or dissimilar methodology now known or hereafter developed.
The use of general descriptive names, registered names, trademarks, service marks, etc. in this publication does not imply, even in the absence of a specific statement, that such names are exempt from the relevant protective laws and regulations and therefore free for general use.
The publisher, the authors and the editors are safe to assume that the advice and information in this book are believed to be true and accurate at the date of publication. Neither the publisher nor the authors or the editors give a warranty, express or implied, with respect to the material contained herein or for any errors or omissions that may have been made. The publisher remains neutral with regard to jurisdictional claims in published maps and institutional affiliations.

Printed on acid-free paper

This Springer imprint is published by Springer Nature
The registered company is Springer International Publishing AG
The registered company address is: Gewerbestrasse 11, 6330 Cham, Switzerland

Preface

"If one tries to discover something new in science, one should avoid following well-worn paths. Along them there is nothing to be gained. Instead, one has to leave the road and to cross untrodden spaces".

This advice was given in the Age of Enlightenment by the German physicist Georg Christoph Lichtenberg. It contains a valuable nugget of truth about the nature of scientific progress. In fact, the ascendance of science in modern times has only been possible because science, again and again, has crossed borderlines to abandon its well-trodden paths.

Such border crossings are also the subject of this book. It asks the provocative question of the computability of the world. This is done by reference to the all-embracing phenomena that—in the traditional view—seem to lie beyond rigorous scientific computability: life, time and history. Behind this hides the deeper question of whether the complex phenomena of our world, which are often loaded with sense and meaning, can become the subject of an abstract science that is based on experiments and mathematics.

How far can the existence and the colourful variety of living beings, the unique character of information and language, the compelling beauty of Nature, the mysterious essence of temporality and the singular historicity of the world be explained and understood within the framework of the exact sciences—which by their very nature are designed to investigate comparatively simple phenomena of our world? It is obvious that the philosophical question of the possibilities and limits of human perception and knowledge will always resonate throughout this set of problems.

The chapters of this book are based upon lectures that I have given on various occasions for interdisciplinary audiences. Each chapter has been thoroughly revised and attuned to the others, with further material added, to make up (I hope) a unified whole. Nevertheless, one has to regard this book, to use another phrase of Lichtenberg, as an "experiment with ideas". As is characteristic of any true experiment in science, the outcome will ultimately only teach us which of our ideas was wrong. This also must be seen as the consequence of an open-minded science, which is permanently changing and progressing.

Munich, Germany   Bernd-Olaf Küppers
January 2012

Note on the English Edition

The questions dealt with in this book were the subject of a series of lectures that I have given over the last 30 years, mainly in Germany. In order to make these lectures accessible to a wider audience, the publisher has encouraged me to prepare an English translation and to revise and to update them at the same time. Above all, I have felt the need to improve Chaps. 3 and 5, in which I have now included recent advances in the objectification and quantification of semantic information. The research in this particular field seems to me to be of general relevance: not only for the science of information itself, but also for the theoretical foundations of modern biology, which rests largely upon the concept of information.

As far as possible, the references have been rearranged to include only articles and books published in English. Foreign-language quotations are taken verbatim from English translations of the originals wherever an authorized translation is available. All other quotations, mostly from German philosophy and literature of the eighteenth and nineteenth century, have been translated directly from the original in close cooperation with the translator of this book, Paul Woolley, whom I wish to thank for his dedicated work.

Munich, Germany   Bernd-Olaf Küppers
July 2017

Contents

Chapter 1
Is Absolute Knowledge of the World Possible?

In his famous fresco "The School of Athens", Raphael showed the philosophers Plato and Aristotle engaged in disputation. Plato is pointing upwards to the superterrestrial world of ideas and Aristotle is pointing downwards to the manifest world of real objects—a scene symbolic of the towering question of philosophical reasoning: Is there an ideal world of absolute knowledge that is independent of any kind of experience, or is human knowledge always bound to the experience and critical analysis of the real world?

© Springer International Publishing AG 2018   1
B.-O. Küppers, *The Computability of the World*, The Frontiers Collection,
https://doi.org/10.1007/978-3-319-67369-1_1

1.1   The Dream of Parmenides

It seems too good to be true: the idea of inalterable, timeless knowledge; a knowledge that does not depend upon changing observations and experiences, because it has its sole origin in the logic of rational thought; a knowledge that, nonetheless, allows all experience to be ordered within a grand, universal context; a knowledge that is given a priori and is conditioned by nothing but itself. In short: Knowledge that can claim for itself absolute certainty and validity.

Parmenides of Elea and the school of philosophers he founded were probably the first to entertain the vision, more than 2000 years ago, that it might be possible to attain an absolute knowledge of the world. With their effort to recognize the nature of "true being", they initiated an understanding of the world according to which the phenomenal reality is merely the deceptive illusiveness of a true and unchangeable world. This hidden world of true being, Parmenides believed, is only accessible by pure reasoning.

Thus, right at the outset of Western philosophy, the idea coagulated that true knowledge of the world could only be arrived at by following the path of rational thought, a path that is free of any sensual perception, any observation or experience. In this way Parmenides laid the foundations of a philosophy, called "metaphysics", which seeks to uncover the ultimate principles, causes and coherences of the world.

Parmenides claimed that the way towards the truth was revealed to him in a dream by the goddess Dike. This tale, however, served alone the purpose of justifying his assertion that metaphysical insights are superior to all forms of empirical knowledge. Only rational access to reality, Parmenides insisted, will lead to absolute truth, while empirical knowledge will always be deceptive and illusory.

Parmenides' understanding of true being implies that reality is ordered according to rational principles and that true knowledge of this reality culminates in the strict deduction of these principles. Moreover, according to this view, there must not only be an over-arching principle of reason that lies at the bottom of all knowledge, but this principle must also be absolutely certain. Only on this presupposition can any metaphysical knowledge lay claim to being true, once for all time.

Parmenides justified his metaphysics with a statement that is already true because of its linguistic construction: the tautological statement "being *is*". Proceeding from this statement, he developed his doctrine of true being in the form of a didactic poem.

The logic of his arguments can be rendered as following: The statement "being *is*" expresses a timeless truth, one that clearly does not admit variations in tense such as "being *has been*" or "being *will be*". It follows in turn that being will never be extinguished, nor can it originate; it is eternal and unchangeable. Every "movement"—that means in ancient philosophy: every spatial or qualitative change of things—must therefore be regarded as a deceptive image of an empirical world, an image that distorts our view of true and unchangeable being.

Furthermore, the statement "being *is*" also implies its linguistic converse: "non-being *is not*". From this statement Parmenides concluded that there is no empty space. Therefore, being must be seen as a continuum, which has no inner or outer boundaries. This led Parmenides to his final conclusion, that true being is not only unchangeable, but that it has further attributes like that of uniqueness, unity, indivisibility and boundlessness.

In the Eleatic school of philosophy, it was above all the logician Zenon who, with the help of sophisticated paradoxes, sought to prove that the true being put forward by Parmenides really is unchangeable, and that every "movement" within reality which seems to refute this conclusion is merely a delusion of the world that we perceive. To demonstrate this, he skilfully used the idea of the infinite, to show that our impressions of movement, multiplicity, divisibility and the like are deceptions, since they lead to irresolvable logical contradictions.

The abstruse pictures of reality engendered by the doctrine of true being were ultimately based upon the fact that logical truths alone cannot lead to meaningful statements about our experienceable reality. Rather, logical truths apply for all conceivable worlds, so that a concrete reference to our real world has to be established in a further step. For Parmenides this was accomplished in the rider "non-being *is not*", from which he deduced that there is no empty space. However, this is only justified, if being is to be conceived of as something that is "matcrial". Yet precisely this (tacit) assumption—that being can only take on material form—is among the arbitrary constructions that in the end lcad to thc inconsistencies of the Parmenidean world.

It is interesting to note that this weak point in the ontology of Parmenides was already pointed out by Aristotle, who critised Parmenides' simple linguistic usage of the word "being". Instead, Aristotle introduced a distinction between that which *is* ("actuality") and that which *might be* ("potentiality"). This he ultimately raised to a point of departure for his own metaphysics, in which he differs clearly from that of the Eleates.

1.2   In Search of the Archimedean Point of Knowledge

Despite the terminological difficulties with respect to the concept of "true being", which already became apparent in ancient philosophy, the vision that one can arrive at absolute knowledge has persisted right down to modern times. It was sustained, just as Parmenides intended it, by the conviction that such knowledge must be based solely upon rational thought, which is free from the deceptions practised on us by sensation and experience.

The paradigm for the rational reconstruction of reality has been the so-called deductive method, which had been applied in a rigorous way by the mathematician Euclid for the axiomatic foundations of geometry. Ideally—this was the leading thought of modern rationalism—true knowledge of reality should, as in geometry, be deducible in its entirety from a highest, and in itself irrefutable, principle. This principle was assumed not only to be the ultimate reference point of all knowledge, but also to be capable of providing a justification for the claim that the knowledge deduced from it is coherent and true.

The modern search for rational knowledge was mainly influenced by thinkers like Descartes, Spinoza and Leibniz. However, the rational systems which were developed in the 17th and 18th centuries revealed substantial differences regarding the question of how the form and the content of the highest principle of knowledge are to be conceived of.

In the school of German Idealism this question became a matter of violent controversy. Starting directly from Kant's investigation of the basic conditions of gaining knowledge, Fichte, Schelling and Hegel each claimed to possess the key to the perfection of the critical philosophy of Kant. Each was convinced that he had discovered the first principle of knowledge, which could claim absolute truth and which would allow it to place human knowledge on an ultimate foundation. But how were they able to make this remarkable claim? To answer this question we have to involve ourselves, at least a little, in a complex philosophical debate concerning the understanding of the so-called "absolute".

Kant is well known to have pushed his way forward to the foundations of knowledge in his epochal work "Critique of Pure Reason" of 1781. The analysis of the conditions under which knowledge becomes possible led him to the concept of the "transcendental subject" as the source of knowledge, prior to all experience. In his view, the subject's perception of the external world is affected by, as he called it, "things in themselves". Following Kant, "things in themselves" constitute reality intrinsically, that is, independently of how we may experience reality. They make up the cause of the phenomena and their determinedness, but they are themselves not recognizable.

Kant initially appears to have supported a realistic view of the "thing in itself". Later he moves towards an idealistic interpretation, according to which the "thing in itself" is merely a terminological construction, but which nevertheless is a logical necessity for the understanding of the source of human knowledge.

This conception was rejected first by Fichte, in 1794, in his "Doctrine of Science". He argued that the knowledge-engendering function of "things in themselves" leaves knowledge still dependent upon the external world, and that knowledge therefore lacks the property of being unconditional. However, in Fichte's view, unconditionality is an indispensable prerequisite if the knowledge acquired by the transcendental subject is to be absolute and no longer dependent upon changes in external experience.

Fichte therefore set out from the idea that the actions of the transcendental subject must be completely unconditional, that is, caused only by the subject itself. To develop from this idea the first principle of all knowledge, Fichte proceeded in an extremely formal manner. At the top of his system of knowledge he put the formal-logical identity "$A = A$". As a logical truth, the statement "$A = A$" cannot be doubted. It thus fulfils the fundamental philosophical condition that the highest principle of all knowledge of reality must be absolutely true. However, it immediately raises the question of how the abstract statement "$A = A$" can ever be related to the real world. Fichte purported to solve this problem by analysing the meaning of the statement "A = A" under the premise that the highest principle must be unconditional without any restrictions.

To come to the nub of Fichte's considerations: statements are always statements by a subject. Therefore, the statement "A = A" presupposes a subject, the "pure Ego" or "I", that sets "A" in an identity relationship to itself. In this way, however, the formal-logical identity becomes conditional upon the "I", and is no longer unconditional, as is demanded by the highest principle of philosophy—with one exception: the only content of "A" that gives the formal-logical identity the character of the unconditional is the "I" itself. It is the statement of the self-authenticating "I", which poses itself: "I am I" or "I = I". Form and content of the highest philosophical principle are thus conditional upon each other and impart to the highest principle the property of complete unconditionality.

Fichte's further arguments along this line can be summarised as follows: As the "I" opposes a "non–I" to itself, it also posits its object of cognition. Moreover, the differentiated world of objects finally arises from the repeated self-limitation ("negations") of the "I" as the "I" again and again poses itself, within the divisible "I", a divisible "non–I". In this way an increasingly fine network of borderlines arises between the "I" and the "non–I" and of demarcation of the "non–I" by the "I". According to Fichte, the world of objects thus appears as a manifold of demarcations, which arise through the iterative self-demarcation of the autonomous "I".

The radical subjectivism that we encounter here was already in Fichte's time a target of criticism. For example Schelling, initially a loyal follower of Fichte, remarked—not without a certain element of mockery—that the divine works of Plato, Sophocles and other great minds were actually his own, as they—if one construes the subjective idealism consistently—are engendered by him through productive intuition [14, p. 639]. Schelling did indeed recognise Fichte's achievements in restoring the subject-object identity to a central position in philosophical debate, but at the same time he criticised Fichte for relegating identity to the position of a particular feature of subjective consciousness. As a consequence of this, he argued, the identity principle itself would remain "after extraction of all substance from the speculation" as no more than "empty chaff" [12, p. 396].

Fichte's philosophical approach, promoting the perceiving subject to the sole and unconditioned source of knowledge, led inevitably to a contradiction with empirical reality. In respect of its understanding of reality, the subjective idealism of Fichte clearly reveals the same weaknesses, the same loss of reality as did Parmenides' doctrine of true being.

In his "Ideas for a Philosophy of Nature", published in 1797, Schelling attempted to correct this deficiency by first objectifying the subject-object identity and not, as Fichte had done, regarding it as an identity proceeding exclusively from the subject. Moreover, according to Schelling the subject-object identity must be considered as absolute. This means that subject and object are not two separate entities that stand in an identity relationship one to another, but rather that the entire Subjective *is* at the same time the entire Objective, and the entire Objective *is* at the same time the entire Subjective. Only when Subject and Object unambiguously and reciprocally depict one another, one to one, is there—as Schelling believed—no inconsistency between the inner and the outer world.

Unlike Fichte, Schelling regarded the real world as more than just an epiphenomenon of the ideal world. Rather, he saw conceptual and material appearances as two manifestations of one and the same entity, and understood this as an absolute subject–object identity. At the same time he realised that he had to pass beyond the concept of Fichte's "Doctrine of Science" and to regard the "I" as an all-embracing world concept, one that encompassed both the entire Subjective and the entire Objective. Schelling admittedly retained Fichte's idea that the absolute "I" engenders its world of perceptions in an act of free will ("freie Tathandlung"); however, at the same time he interpreted the absolute I as being the highest level of existence of a self-creating Nature, which in the human mind becomes conscious of itself.

Fichte's statement "I = I" thus becomes a self-statement of self-creating Nature: "I = Nature". Consequently, the acts of free will on the part of the absolute "I" are interpreted by Schelling as objective acts of creation of an all-encompassing, autonomous and unconditional activity of Nature, which are raised into the realm of human consciousness. Schelling [12, p. 380] summarises: "Nature is the visible mind, the mind is invisible Nature". This is to be taken as meaning that the perceiving subject can regard itself in Nature as in a mirror. Nature is the visible mind. Conversely, mind is invisible Nature, insofar as mind mirrors Nature at the highest level of its being. Thus, mind in Nature and Nature in mind can contemplate one another.

1.3 Utopian Fallacies

Schelling's philosophy of Nature rests upon the identity of rational and natural principles, which is assumed to be absolute. The aim of his philosophy is to reach knowledge of Nature a priori, which is—, according to the identity relation—, constituted from the principles of rational thought. By definition, this knowledge does not depend upon changeable experience. Consequently it cannot be refuted through experience. At the same time, knowledge a priori is the highest authority to decide over the question of how experience is to be ordered and interpreted. In comparison with this kind of knowledge, the discoveries of empirical research into Nature are only of marginal significance; these only become "knowledge" when their necessity, i.e. predetermined place in the theoretical system of natural philosophy, is recognised. In this sense, Schelling's natural philosophy claims to be a "higher knowledge of Nature", a "new organ for regarding and understanding Nature" [12, p. 394].

In Schelling's system, the task of empirical science is—at best—to verify the principles dictated to it by natural philosophy. On no account could they be disproved: the refutation of these principles would immediately have refuted the principles of reason and, thus, pursued the possibilities of cognition *ad absurdum*.

In fact, the principles of natural philosophy were seen as unchallengeably certain. If empirical results did not accord with them, then the principles remained unchallenged, whereas the empirical observations were taken to be obviously at fault, or incomplete, or deceptive.

Schelling did at least admit that empiricism might occasionally have a heuristic function in the discovery of hitherto unknown phenomena of Nature. However, the constructive rôle thus assigned to empiricism clearly marks it out as a foreign body in the stringent system of a priori epistemology.

There are two possible reasons why Schelling allotted a certain place to empiricism. On the one hand, he was hardly in a position to ignore the enormous and dynamic progress that empirical science had made, and was making, in his time. On the other, Schelling's natural philosophy never really progressed beyond the stage of being a mere epistemological programme, with the result that, to make even moderately concrete statements about the world, he was forced to resort to observation. This, too, shows that pure thought, free from any relation to experience, is unable to attain any recognition of experienceable realities.

Whatever Schelling's motives may have been in raising the status of empiricism, there is little doubt of their incompatibility with his philosophical programme of Nature. Even if Schelling occasionally re-interpreted experience as a priori knowledge that only appears to be a knowledge a posteriori as long as its transcendental-logical roots remain unrecognised, he kept a firm grip on the dogmatic epistemological claim of natural philosophy that its conclusions, as long as they were correctly inferred, cannot be falsified by experience but only verified. For him, the function of empiricism was at most to guide discovery, never to provide it. On the contrary: real discovery could only be made by natural philosophy.

A further aspect of Schelling's epistemology should be emphasized. In accordance with the identity principle, the ideal and the real together make up a whole that cannot be transcended. The whole is at the same time an allegory for the absolute, which however only reveals itself in the dichotomous form—that is, in ideal and real essence—to the subject. However, the absolute, when it "expands" into the ideal and the real, must not lead out of the absolute. As the absolute, it must always remain identical with itself in its entire absoluteness.

From this there inevitably emerges a picture of the holographic character of the world, which is the same as saying that "every piece of matter must in itself bear the imprint of the entire Universe" [13, p. 413]. According to Schelling, this is not least applicable to the relationship between Nature and organism—with the consequence that Nature is to be regarded as a universal organism, which is arranged according to the same principles as an individual organism and vice versa. Admittedly, organic matter, as an expression of the absolute, is not a static structure but, as Schelling emphasised again and again, a process. Only in this way was Schelling able to arrive at a self-consistent picture of Nature as pure natural activity that, in "infinite productivity", engenders all natural objects out of itself.

Natural philosophy and empirical research into Nature are thus concerned with two fundamentally different objects of knowledge. One is concerned with "Nature as a subject" and the other with "Nature as an object". "Nature as a subject" is a metaphor for the infinite productivity of Nature ("natura naturans"). It is downright natural dynamics. Its driving forces are the creatively acting natural principles, the discovery of which is the task of natural philosophy. "Nature as an object", in contrast, is the productivity of Nature as made manifest in her products ("natura naturata"). These products are in themselves finite and appear as a terminated network of actions, the elucidation of which is the task of empirical research into Nature. However, to avoid the conceptional separation of Nature into two forms, Schelling employed an artifice. According to this, the productivity of Nature is not really extinguished in its products; rather, it still remains active with a force of production that, however, is infinitely delayed. As already encountered in the philosophy of the Eleates, the concept of the infinite again must be invoked in order to save the consistency of the epistemological model.

The organismic conception of Nature, as developed in detail by Schelling in his book "On the World Soul" entails an important consequence for the method of gaining knowledge. As the parts can be explained on the basis of the whole, Schelling argued, the whole—that is, the organismic character of Nature—is the basis for explanation of all natural phenomena from the highest levels of complexity of matter down to its simplest parts. In this way, the direction of explanation encountered in modern sciences, which leads from the simple to the complex, was turned upside-down. According to Schelling the organism is not to be explained on the basis of its material building blocks, but rather these building blocks must be explained on the basis of the overall picture of the organism. The organismic conception of Nature was thus given precedence over the mechanistic one.

In summary, we can say that Schelling's philosophy of Nature ran counter to today's scientific method in two important respects: (a) Theory occupies a more important place than empiricism. Claims to truth need not stand the test of experience; they are exclusively derived from logical reasoning. In short: Knowledge a priori is given precedence over knowledge a posteriori. (b) The research strategy propagated by Descartes, Newton and others, according to which one should proceed from the simple to the complex, from the part to the whole, from the cause to the effect, is turned by Schelling into its opposite. The analytical method, based upon dismantling, abstraction and simplification, is discarded—or at least diminished in importance—in favour of a holistic method (for a detailed criticism of this kind of philosophy see [5]).

Schelling's philosophy of Nature was not the only attempt made in the 19th century to dictate to empirical research which way it had to go. Hegel, Eschenmayer, Steffens and others likewise developed their own ideas of natural philosophy. However, none of these had anything like the effect that was attained by Schelling's approach. Schelling, with his metaphysical, speculative understanding of Nature, was the only philosopher to succeed in initiating a counter-movement to the mechanistic sciences and to found the so-called "romantic" philosophy of Nature, a world-view that even today finds numerous adherents.

Schelling's philosophy emerged from the sober logic of the rationalistic perception of reality. One may therefore ask how this philosophy ever acquired the attribute "romantic". To answer this is no easy matter, as—to start with—the term "romantic" does not have a clear meaning. It is one of those elastic words whose meaning is only clear within the particular context in which it is used.

In the present case, this context is as follows: At the beginning of the 19th century, Schelling's philosophy offered an alternative to the then predominant mechanistic view of Nature. It presented the mechanistic view of reality as a constricted perspective of a world which in actual fact is a complex whole. Moreover, Schelling's conception of Nature as an all-embracing organism appeared to correspond perfectly to the romantic ideal of an organic, indivisible unity of Man and Nature. In this way his philosophy took on a constitutive rôle for the romantic understanding of the world. Nevertheless, the organismic conception of Nature propagated by Schelling led directly into the fog of a romantic transfiguration of Nature, in which, even today, adherents of a romantic understanding of Nature appear to be straying about.

The vagueness of this view of Nature can easily be illustrated by examination of its concept of the whole. According to the organismic view, the phenomena of Nature inherently make up a unified whole and must be recognised from the per-spective of this unity. For this, even in our times, again and again the idea of a holistic method for the understanding of the organism is propagated—a method believed to be in contrast to causal-analytical thinking. However, the idea of an irreducible whole is anything but transparent. It cannot even be explicated mean-ingfully, let alone be determined by analytical thought. In the end, all that remains is the tautological conception of "the whole" as some kind of "whole".

This problem refers back to the problem of the absolute. In fact, the concept of the whole was introduced by Schelling precisely for the purpose of giving appro-priate expression to the absolute. The absolute appears, however, persistently to resist analysis, because it—being unconditional – completely eludes any relation-ship of conditionedness, not least that of reflection. The unconditional, as a syn-onym for the absolute, cannot be subjected to the conditions of consideration. There is no Archimedean point outside the absolute, one from which the absolute might be determined. The absolute as such is indeterminate. For exactly this reason, one encounters repeatedly in Schelling's philosophy empty formulas according to which we are invited to think of the absolute first and foremost as "sheer abso-luteness". And in the passages where Schelling finally does undertake the endeavour to encircle the absolute conceptually, his thought dissolves into poetry.

The addendum to the Introduction of Schelling's "Ideas of a Philosophy of Nature", in which he repeatedly attempts to express the inexpressible, is rich in morsels of poetic word-creation and pictorial comparison that exhaust themselves in nebulous abstraction. We read, for example, that the absolute is "enclosed and wrapped up into itself", or that the absolute "is born out of the night of its being into the day". There Schelling speaks of the "æther of absolute ideality" and the "mystery of Nature" [12, p. 387 ff.].

The poetic language that Schelling makes extensive use of is clearly the inevi-table accompaniment of a philosophy in which human thinking perpetually seeks to transcend itself. Only thinking about the absolute can be reflected in it, but not the absolute itself. Consequently there arises an unsolvable intellectual problem, an *aporia*, regarding the absolute, which Schelling tried to circumvent by introducing the concept of intellectual intuition ("intellektuale Anschauung"). This means the contemplative act of self-ascertainment of the absolute by the introspective self-consideration of the absolute. Thus, intellectual intuition appears like an inwardly inverted Archimedean point, from out of which the absolute was supposed to be made comprehensible.

Schelling generalised the concept of intellectual intuition (which had already been used by Fichte) and abstracted it from the beholder and thus, as he put it, only considers the purely objective part of this act. Nevertheless, in this way a quasi-meditative, almost occult element crept into his natural philosophy, and remained permanently stuck to it like an annoying vermicular appendage.

Despite its romantic exaggeration by the organismic view of Nature, Schelling's philosophy is nonetheless, deep down, even more mechanistic than the mechanistic sciences that he criticised so violently. It is true that he attempted to replace the causal-analytical view, according to which the world can be described as linear chains of cause and effect, by an organismic approach with cyclic cause-and-effect relationships. At the same time, however, everything was subjected to the highly mechanistic logic of deductive philosophy. That in this way the experience and the consideration of Nature ultimately came under the wheels, was an objection that even Goethe raised against the proponents of romantic philosophy of Nature.

Goethe was initially sympathetic to the aims of these philosophers and gave them active encouragement by supporting, for example, Schelling's appointment to Jena University; later, however, he turned away, "shaking his head", from their "dark", "ambiguous" and "hollow" talk, which he felt was "in the manner of prophets" [3, p. 483 f.]. He went so far as to see in their speculative philosophy an "ugly mask", a thing "highly fantastical and at once dangerous", because here "the formulae of mathematics pure and applied, of astronomy, of cosmology, of geology, of physics, of chemistry, of natural history, of morals, religion and mysticism [...] were all kneaded together into a mass of metaphysical speech", with the consequence "that they substitute the symbol which suggests an approximation, for the real matter; that they create an implied external relationship to an internal one. Thus, instead of exposing the matter they lose themselves in metaphoric speech" [3, p. 484].

More so: even the closest comrades-in-arms of the Jena Romantics' Circle, such as Friedrich Schlegel and Johann Wilhelm Ritter, criticised the notion that pure speculation, unaided by any experience, could provide the basis for a any profound knowledge about the world. "Schelling's philosophy of Nature", concluded Schlegel, "will inevitably invoke strong contradiction from the crass empiricism that it had hoped to destroy" [15, p. 50]. And Ritter insisted that "pure experience [...is...] the only legitimate device to allow the attainment of pure theory" [11, p. 122]. We shall, according to Ritter, "approach imperceptibly the true theory, without searching for it—we shall find it by observing what really happens, for what more do we desire of the a theory than that it tells us what is really happening?" [11, p. 121].

Ritter rejected the claim of Schelling's philosophy to deliver a theoretical foundation a priori of all natural phenomena. But at the same time he moved towards this philosophy because he made its speculative theses into guidelines of his experimental studies. Other scientists, such as Hans Christian Ørsted, Lorenz Oken and Carl Gustav Carus, moved in a similar direction. From this trend there finally emerged an approach to research, which its proponents regarded as "romantic study of Nature" (see above). Not least, they were engaged in finding an experimental demonstration of the mysterious force that, as was assumed at the time, permeates and connects all organic and inorganic matter.

Toward the end of the 18th century, the field of choice for experiments in romantic research into Nature was above all the phenomena of electrochemistry. At the centre of Ritter's studies, for example, were the observations made by Luigi Galvani that a suitably prepared frog's leg could be set in motion by electric currents. Everything seemed to point to galvanic electricity as the key to understanding living matter. Moreover, this fitted in with the—at the time—highly popular conception of polarity and of enhancement (another idea popularised by Goethe) as the "two great driving wheels of all Nature" [2, p. 48].

In his writing "On the World-Soul" Schelling believed not only that he could provide a rigorous philosophical foundation for this conception, but also that his ideas had found impressive confirmation in Galvani's observations. For the experimenter Ritter, the phenomena of polarity and enhancement became the central guiding principle of his research. His urge to discover rose to the degree of an obsession, as he began to conduct electrochemical experiments on his own body. The fact that Ritter's self-experimentation ultimately brought about his death can be regarded as a macabre climax of this period of dogmatic understanding of Nature, in which the borders between self-knowledge and knowledge of Nature, between the human body and the corpus of Nature, were blindly negated (see [1]).

1.4 From the Absolute to the Relative

For Schelling, knowledge of Nature consisted in reconstruction of the self-construction of Nature. The concepts of natural philosophy were thus regarded as necessarily true, and considered to be immune from refutation by experience. In this they differed fundamentally from the theories of empirical research into Nature; these have only a relative validity and are always subject to critical examination by experience.

In the claim of natural philosophy to represent a "higher" knowledge of Nature, we recognise once more an idea of ancient metaphysics that originated with Parmenides: the idea that phenomenological reality, upon which empiricism is founded, is only the surface of a deeper-lying, true reality. In accordance with this, Schelling conceived of Nature in a double sense: "Nature as object" and "Nature as subject".

He argued in the following way: As empirical research into Nature is directed exclusively at "Nature as object", its discoveries are all only provisional. Natural philosophy, in contrast, is directed towards "Nature as subject". Its discoveries are valid once and for all time. With its help, however, empirical knowledge can be ordered and dovetailed into the general system of knowledge. In this manner the theorems of natural philosophy form the immovable co-ordinates of all knowledge. Whoever knows these co-ordinates, Schelling claimed, is in possession of a priori knowledge, which enables him to study systematically both Nature and, in doing so, reality itself.

It was the absolute certainty with which knowledge seemed to be acquirable that made the speculative philosophy of Schelling appear so attractive to the romantic searchers of Nature. Moreover, they believed that the instrument of analogy put them in possession of a universal procedure for discovering Nature's hidden phenomena.

However, the actual contribution of romantic research into Nature was scant. Even the important discoveries made by Ritter (ultraviolet radiation) and Ørsted (electromagnetism) hardly affect this balance. It is true that both were convinced they owed their discoveries to the principles of natural philosophy, but in retrospect that belief appears highly questionable. It seems much more to confirm the well-known fact that, with a goodly portion of chance, trail-blazing discoveries are sometimes made on the basis of completely wrong ideas. Thus, the discoveries of Ritter and Ørsted rest not on any necessary connection between experimental findings and the theorems of natural philosophy, but simply upon the fortuitous coincidence of the two. Aside from a superficial construction of analogies, the ideas of natural philosophy produced no logically compelling argument from which experimental research into Nature could have been inferred. Apparently, the romantic researchers achieved their successes by the same process as the blind man who seeks an object with the help of a lamp and, quite by chance, finds it.

The physical chemist and philosopher Wilhelm Ostwald likewise saw no way of explaining the successes of romantic research into Nature except by appeal to the "complete inability to shy away from the absurd, of which the philosophers of Nature make such copious use". Only in that way, he wrote, were the romantic researchers in a position to "find analogies that were in fact present, while their contemporaries missed sight of them because of their unfamiliar form" [9, p. 8].

Even in Schelling's time, the speculative philosophy of Nature was the object of harsh criticism from almost all the leading mathematicians and scientists. For Wöhler, Liebig, Herschel, Virchow, Helmholtz and others, as Carl Friedrich Gauß put it, "their hair stood on end at the thought of Schelling, Hegel, Nees von Esenbeck and their consorts" [10, p. 337].

This annihilating criticism is very easily understood when one makes the effort to review, for example, Hegel's Philosophy of Nature. Hegel not only wrote off Newton's theories of light out of hand as "nonsense" [4, p. 254]; even the most elementary knowledge of contemporary science, according to which lightning consists of electric discharges or water of hydrogen and oxygen, were challenged. "The healthy person", according to Hegel, apparently projecting from himself onto others, "does not believe in explanations of that kind" [4, p. 146]. Instead, Hegel constructed worlds in which solar eclipses influence chemical processes, in which magnetic phenomena are dependent upon daylight, in which the solar system is an organism, in which heavenly bodies do not appear at certain places because, according to the dogmatic character of the speculative philosophy, nothing can be the case that is not allowed to be the case.

Hegel's self-styled discoveries were wrapped up in incomprehensible, empty and downright absurd sentences as the following passages *pars pro toto* demonstrate: "The real totality of the body, as the unbounded process that individuality determines toward particularity or finitude and the same likewise negates and withdraws into itself, at the end of the process re-forms itself as at the beginning, is therewith an upward step into the first ideality of Nature, so that it has become a *fulfilled* and essentially self-relating *negative* unity, a *self-referential* and *subjective* one" [4, p. 337], or "The resolution of tension as rain is the reduction of the earth to neutrality, a sinking into unresisting indifference" [4, p. 152]. With Hegel's "derivation" of the order of the planets (which, scarcely promulgated, was immediately refuted by observation) the speculative philosophy of Nature finally began to verge upon the ridiculous. The polemics with which Schelling had once castigated the allegedly "blind and unimaginative kind of natural science which has taken hold since the deterioration of philosophy and physics at the hand of Bacon, Boyle and Newton" [12, p. 394], now returned like a boomerang and hit the philosophy of Nature squarely. For Liebig, this kind of philosophy was a "dead skeleton, stuffed with straw and painted with make-up", "the pestilence, the Black Death of our century" [8, pp. 23–29]. Wöhler saw in its promulgators "fools and charlatans", who did not even themselves believe the "hot-headed stuff" that they preached [17, p. 39]. Schleiden lost his patience over "the crazy ideas of these caricatures of philosophers", whom he brushed off as "sham philosophers" [16, p. 36].

There was just one single, initially almost impregnable bastion of natural philosophy: romantic medicine. The human organism, with its unimaginable material complexity, organised as it is in a seemingly infinite number of cause-and-effect loops, appeared completely resistant to any attempt at a mechanistic analysis. Thus, in the nature of things, the organismic research paradigm seemed to offer itself as a valid way to regard the organism, one that would not so quickly be caught up with by mechanistically oriented research. Romantic medicine thus succeeded in establishing itself alongside academic medicine as an independent paradigm for medical research. Even today, holistic medicine and its more esoteric fringe doctrines carry significant weight in the shape of so-called complementary medicine.

This is not the place to trace the entire history of the impact and public perception of the romantic philosophy of Nature with all its ramifications—not least because this philosophy had no lasting effect upon the natural sciences, and because its impact and public perception had already exhausted themselves in the polemic debates between this kind of philosophy and the exact science. "That shallow twaddle", wrote Schleiden [16, p. 35], the founder of cell biology, "had no influence whatever upon astronomy and mathematical physics, but for a while it confused the life sciences; however, once they had grown out of these penurious ideas, they rightly branded Schelling as a 'mystagogue', and he disappeared in the smoke of his own mythological philosophy".

Nowhere is the unbridgeable chasm between metaphysical and empirical knowledge of the world more clearly visible than in a direct comparison with today's conception of science. While the idea of the absolute is central and indisputable in the metaphysical and speculative philosophy, it is precisely a characteristic of the modern sciences—experientially and positivistically oriented as they are—that they have rigorously eliminated the idea of the absolute. It is only against this background that one can really comprehend the development of the modern sciences.

Physics, for example, owes its enormous progress over the past hundred years above all to the fact that it has repeatedly abandoned absolute concepts in favour of relative ones. Well-known examples of this are the relativity of space and time in Einstein's theory, and of the concept of 'object' in quantum theory. But in biology, too, a comparable process seems to be taking place. Here it is the basic concept of information that is increasingly being recognized as a relative one, with important consequences for the understanding of the origin and evolution of life (see [6] and [7]).

The abandonment of the absolute in favour of the relative is precisely the force behind the strongest impulses toward the development of modern science. This is not least true of the question of knowledge itself. The natural philosophy of the 18th and 19th century may have regarded relative knowledge as imperfect and incomplete, to be overcome in the striving for absolute knowledge; but for empiricist–positivistic research it is precisely the relative character of knowledge that guarantees progress. Only knowledge that can be corrected is able to lead us out of epistemological impasses. Only openness toward the lessons of experience can provide the basis of genuine progress in the acquisition of knowledge. And this progress does not consist in the accumulation of incontrovertible truths. On the contrary, paradoxical as it may sound, progress in scientific knowledge becomes more and more precise, because we learn more and more about what is *wrong*.

For the modern, empirical sciences, which base their claims to truth and validity upon the corrective power of experience, the dream of absolute knowledge must seem more like a nightmare. The idea that one might be able to derive from a single fundamental principle the rich, living content of reality—without a single reference to observation or experience—stands diametrically opposed to today's scientific concept of reality. The mere analysis of the meaning of terms like that of the absolute, from which the romantic philosophy of Nature drew its sole support, can never replace the knowledge-guiding and knowledge-founding function of experience.

The systematic subjugation of experience to ratiocination was, in Ostwald's words, "the error that the philosopher of Nature made, and one that we must avoid at any price. They tried to derive experience from thought; we shall do the opposite and let our thought be governed everywhere by experience" [9, p. 7]. In this sense, the natural sciences have decided to follow consistently the empiricist-positivistic path, while the romantic philosophy of Nature, snarled up in the aprioristic constraints of its metaphysical world view, was crushed by the scientific developments in the 19th and 20th centuries.

Nonetheless, attempts have repeatedly been made to reanimate romantic philosophy of Nature and to instrumentalise it in the cause of an antimechanistic view of the world. These attempts have not been taken very seriously in the exact sciences: too deep is the chasm between natural science, which only admits knowledge as being of a hypothetical nature, and the various branches of philosophical world foundations, which never question their own knowledge because they are convinced of being in possession of final understanding and ultimate truth. However, only one thing is certain: The metaphysical dream of absolute knowledge will for ever be a dream unfulfilled.

References

1. Daiber, J.: Experimentalphysik des Geistes. Vandenhoeck & Ruprecht, Göttingen (2001)
2. Goethe, J.W. von: Werke. Hamburger Ausgabe in 14 Bänden, Bd. 13. Beck, München (1981)
3. Goethe, J.W. von: Goethes Briefe und Briefe an Goethe, Hamburger Ausgabe in 6 Bänden, Bd. 4. Beck, München ^{3}(1988)
4. Hegel, G. W. F.: Werke, Bd. 9. Suhrkamp, Frankfurt am Main 1986
5. Küppers, B.-O.: Natur als Organismus. Klostermann, Frankfurt am Main (1992)
6. Küppers, B.-O.: The context-dependence of biological information. In: Kornwachs, K., Jacoby, K. (eds.): Information, pp. 137–145. Akademie Verlag, Berlin (1996)
7. Küppers, B.-O.: The nucleation of semantic information in prebiotic matter. In: Domingo E., Schuster P. (eds.): Quasispecies: From Theory to Experimental Systems, pp. 23–42. Springer International Publishing, Switzerland (2016)
8. Liebig, J. von: Über das Studium der Naturwissenschaften und über den Zustand der Chemie in Preußen. Vieweg, Braunschweig (1840)
9. Ostwald, W.: Naturphilosophie. Veit & Comp., Leipzig ^{3}(1905)
10. Peters, C. A. F. (ed.): Briefwechsel zwischen C. F. Gauss und H. C. Schumacher, Bd. 4. Esch, Altona (1862)
11. Ritter, J.W.: Beyträge zur nähern Kenntniß des Galvanismus und der Resultate seiner Untersuchung, Bd. 1. Frommann, Jena (1800)
12. Schelling, F. W. J.: Ideen zu einer Philosophie der Natur als Einleitung zum Studium dieser Wissenschaft. In: Schriften von 1794–1798. Wissenschaftliche Buchgesellschaft, Darmstadt (1980)
13. Schelling, F. W. J.: Von der Weltseele. In: Schriften von 1794–1798. Wissenschaftliche Buchgesellschaft, Darmstadt (1980)
14. Schelling, F. W. J.: Über den wahren Begriff der Naturphilosophie und ihre richtige Art Probleme aufzulösen. In: Schriften von 1799–1801. Wissenschaftliche Buchgesellschaft, Darmstadt (1990)
15. Schlegel, F.: Europa. Eine Zeitschrift, 1. Bd., 1. Stück. Frankfurt (1803)
16. Schleiden, M.J.: Über den Materialismus der neueren deutschen Naturwissenschaft, sein Wesen und seine Geschichte. Engelmann, Leipzig (1863)
17. Wöhler, F.: Briefwechsel zwischen J. Berzelius und F. Wöhler, Bd. 1. Engelmann, Leipzig (1901)

Chapter 2
Are There Unsolvable World Enigmas?

With the help of huge machines, science endeavours to wrest the last secrets from Nature. The photograph shows the large accelerator at the European nuclear research centre CERN, which is designed to simulate processes that presumably took place at the origin of the universe. Can we hope to understand some day the existence of the world, or does the origin of all things confront us with an unsolvable problem?

© Springer International Publishing AG 2018

B.-O. Küppers, *The Computability of the World*, The Frontiers Collection,
https://doi.org/10.1007/978-3-319-67369-1_2

2.1   The "Ignorabimus" Controversy

The physiologist Emil Du Bois-Reymond was one of the great scientific personalities of the 19th century. He is best known to specialists for his electrophysiological studies of muscle and nerve excitation. Beyond that, he was also passionately involved in debating the broad social and philosophical questions of his time.

In conformity with the scientific *Zeitgeist* of the late 19th century, Du Bois-Reymond was a committed adherent to the mechanistic world-view, according to which the ultimate goal of knowledge of Nature was to demonstrate that all events in the material world could be traced back to movements of atoms governed by natural laws. Every natural process—that was the credo of the mechanistic maxim—rests ultimately upon the mechanics of atoms.

For Du Bois-Reymond, the rise of the natural sciences in the 19th century represented the epitome of cultural progress. This led him to formulate the provocative statement that "the real history of mankind is the history of the natural sciences" [2, p. 134]. It must therefore have appeared all the more contradictory when at the same time Du Bois-Reymond claimed, with almost missionary zeal, that certain world problems could never be solved by the natural sciences. Here, he was referring especially to humanistic issues. These, he maintained, could never be explained within the framework of a mechanistic view of Nature, because feelings, emotions and thoughts were fundamentally incapable of being naturalised and therefore could not be explained in terms of the mechanical laws of non-living matter.

By propagating this view, Du Bois-Reymond contradicted an idealised view of mechanistic Nature that went back to the mathematician Pierre Simon de Laplace. At the height of the mechanistic age, Laplace had conjured up the vision of a universal mind that, with the help of the mechanical laws and complete knowledge of the momentary state of the world, would be able to calculate all its future and past states. Nothing remained hidden from Laplace's "universal mind" in the transparent world of mechanisms. Even human thoughts, emotions, actions and suchlike were calculable in principle, as—according to the mechanistic world-view —all mental phenomena were merely particular expressions of material properties.

For Du Bois-Reymond this radical view of the mechanistic maxim was untenable. He was deeply convinced that mental phenomena were immaterial in nature, and for this reason were fundamentally inaccessible to any mechanistic analysis. The attainments of the human mind, he asserted, could in reality only reach a level that was a pale reflection of Laplace's Universal Mind. Notwithstanding, even this was subject to the same limitations as we are, so that the unsolvable riddles that tantalised human thinking would also be impenetrable for the Universal Mind.

In a famous speech, given to the Convocation of German Naturalists and Physicians, Du Bois-Reymond listed "seven world enigmas", which he considered to be unsolvable [1]:

1. The nature of force and matter,
2. The origin of movement,
3. The origin of life,
4. The apparently purposeful, planned and goal-oriented organisation of Nature,
5. The origin of simple sensory perception,
6. Rational thought and the origin of the associated language,
7. The question of free will.

Some of these issues Du Bois-Reymond regarded as "transcendent" and therefore unsolvable. Among these, in his view, was the question of the origin of movement. Other problems, such as the origin of life, he regarded as solvable, but only insofar as matter had already begun to adopt movement. However, Du Bois-Reymond saw the seven world enigmas as progressing upwards, each dependent upon its predecessors in the list; consequently, the sum of all these seems to constitute a coherent, fundamentally unsolvable complex of problems. In that sense, he ended his speech on the limitations of possible knowledge of Nature with the words "Ignoramus et ignorabimus" (we do not know, we shall not know).

With his "Ignorabimus" speech, Du Bois-Reymond set off a violent scientific controversy. It continued to be fought out, sometimes polemically, by the supporters and opponents of his theses. Du Bois-Reymond himself was completely aware that his "Ignorabimus" was ultimately "Pyrrhonism[1] in new clothing" and would inevitably evoke contradiction from the naturalists [1, p. 6]. In fact, Du Bois-Reymond did not need to wait long for this. Soon, his most powerful adversary emerged in the no less famous scientist Ernst Haeckel. The latter was professor of zoology in Jena and had already emerged as a vehement supporter and promulgator of the Darwinian theory of evolution. The central element in Darwin's thinking, according to which the evolution of life rested on the mechanism of natural selection, appeared to confirm to the fullest extent the materialistic world picture favoured by Haeckel.

On the basis of a materialistic world-view that was gaining increasing acceptance, Haeckel stood for a monistic doctrine of Nature. According to this, even intellectual phenomena are nothing more than material processes, unrestrictedly accessible to mechanistic explanations of the kind that Darwin had put forward. Haeckel took the view that the aprioristic forms of intuition, which according to Kant were prerequisites for the possibility of any cognition, could be interpreted and explained as a posteriori elements of phylogenesis. By so doing, Haeckel anticipated an idea that the ethologist Konrad Lorenz [11] took up in the mid-twentieth century and developed into the evolutionary theory of cognitive mechanisms in animals and humans.

Haeckel opposed vigorously the dualistic view of Mind and Matter. Consequently, he challenged the existence of unsolvable "world enigmas" of which

---

[1]Pyrrhonism, also termed scepticism, goes back to the ancient philosopher Pyrrho of Elis and refers to a philosophical movement that cast fundamental doubt upon the possibility of a true knowledge of reality.

Du Bois-Reymond had spoken. Except for the problem of free will, which he regarded as a dogma because it was (in his view) illusory and did not in fact exist, he saw the other world problems either as solved within the framework of the materialistic conception of Nature, or else as solvable in principle.

As a rhetorician, Haeckel was by no means second to the "orator of Berlin", as he decried his opponent. Haeckel, too, knew exactly how to popularise his world-view, and he used this as an opportunity to exacerbate the "Ignorabimus" controversy into a battle of cultures, between the monistic and dualistic world-views. Du Bois-Reymond, on his part, mocked the "prophet from Jena" and adhered tenaciously to his "Ignoramus et ignorabimus" slogan, which for him—together with his seven world enigmas—was an "unchangeable and inflexible verdict" [1, p. 51].

The "Ignorabimus" thesis made up the strongest countercurrent to the optimistic attitude to knowledge taken by the natural scientists of the 19th century, which is why it provoked such violent criticism from Haeckel. The modesty of "Ignorabimus", according to Haeckel, is a false modesty; in reality, he asserted, it is an expression of presumptuousness, as it claims to lay down limits to knowledge of the natural world that apply for all time, and to raise ignorance to the status of an absolute truth.

Yet, Haeckel proceeded to pour more oil onto the fire: "This seemingly humble but really audacious '*Ignorabimus*' is the '*Ignoratis*' of the infallible Vatican and the 'black international' which it leads; that mischievous host, against which the modern civilzed state has now at last begun in earnest the 'struggle for culture'. In this spiritual warfare, which now moves all thinking humanity, and which prepares the way for a future existence more worthy of man, spiritual freedom and truth, reason and culture, evolution and progress stand on the one side, marshalled under the bright banner of science; on the other side, marshalled under the black flag of hierarchy, stand spiritual servitude and falsehood, want of reason and barbarism, superstition and retrogression." [6, p. xxii f.]. Du Bois-Reymond, in turn, felt himself "denounced as belonging to a black band of robbers" by such utterances and saw in them the proof of "how close to one another despotism and extreme radicalism dwell" [3, p. 72].

The long-lasting effect of the "Ignorabimus" controversy in science is revealed clearly in a radio broadcast given in 1930 by the mathematician David Hilbert, who was one of the leading intellectual pioneers of his time. Hilbert ended his talk with the words: "We must not believe those who today, with philosophical bearing and in a tone of superiority, prophesy the downfall of culture and admire themselves as adherents of the Ignorabimus. For us there is no 'Ignorabimus', and neither, in my view, is there any room for it in natural science. In place of the folly of the 'Ignorabimus', our watchword is: We must know, we shall know." [8]. In the same spirit, Hilbert had challenged his colleagues at the beginning of the 20th century with a legendary list of fundamental mathematical problems, the solving of which he considered to be of highest priority.[2] Hoewever, very soon after Hilbert's

---

[2]At the International Congress of Mathematicians of 1900, held in Paris, Hilbert formulated with exemplary exactitude a list of fundamental problems that he claimed to be solved. In 1928 he extended this list.

broadcast, the logician Kurt Gödel proved that one of these problems, known as Hilbert's 23rd problem, was principally unsolvable.

Today we are well aware that there indeed are unsolvable problems in science, or at least in mathematics. In retrospect, it seems like an irony of fate that Hilbert, of all people, set out a problem that was finally found to be unsolvable. Nevertheless, Hilbert's maxim "We must know, we shall know" became the unchallenged *Leitmotiv* of modern science, which persistently ignores the "Ignorabimus" and resolutely endeavours to comprehend the incomprehensible, to calculate the incalculable and to measure the immeasurable. And, in the same spirit, the exact sciences are perpetually pushing at, with the aim of overcoming, the various boundaries of our knowledge of the world.

2.2   Crossing Boundaries in Science

All areas of human life are criss-crossed by boundaries. These clearly constitute—irrespective of any question of whether they are real or imaginary borders—a rich picture of reality. It is only on the basis of the manifold boundaries between natural and artificial, living and non-living, law-like and random, simple and complex, regular and unique, and many other pairs of opposites, that we encounter reality in differentiated forms. However, human thought has always felt the challenge to question these boundaries and, where possible, to cross them. This is true not least of the—presumably—most fundamental boundary, that of human self-awareness: the abyss between mind and matter, on which, in its time, the "Ignorabimus" controversy flared up.

At first glance, it would appear that the differentiated richness of reality must become lost in precisely the same measure as the acquisition of knowledge moves or abolishes borders. Yet the apparent loss is compensated for by the gain in unity of human knowledge. This is in turn a prerequisite for a coherent picture of reality, one that is free of internal contradictions.

The human drive towards discovery and thirst for knowledge are ultimately impelled, to a large extent, by perpetual reflection on the various boundaries that cross the material and non-material world. However, what kind of boundaries do we encounter in science? The first that springs to mind is a conceptional boundary in mathematics that was discovered by Gödel in the 1930s (see above). With the proof of his so-called "incompleteness theorem" Gödel had demonstrated the unsolvability of Hilbert's 23rd problem, and had revealed the fundamental unrealisability of Hilbert's idea that the path of mathematical proof could in principle be completely automated.

Indeed, Hilbert's programme aimed at placing mathematics on a new foundation, and had for a very long time pursued the goal of constructing a completely formalised system in which all true theorems of mathematics could in principle be proven by applying a standardised procedure. When Gödel managed to prove with mathematical rigour that any such formal system is for fundamental reasons incomplete, he showed once for all that mathematical intuition and insight are fundamentally incapable of being replaced by formalised thought. In other words: an ultimate demonstration of the truth of any knowledge by an automatic proof is not possible, because truth encompasses more then mere provability. In this way, Gödel had uncovered a boundary in formal thought, one that constitutes a genuine and insurmountable barrier.

At the centre of Hilbert's formalistic programme there already lay, in essence, the notion of an idealised computer, a so-called "Turing machine" – even though this idea only took on form some decades later, when it was set out and developed into a fully-fledged theory by the mathematician Alan Turing.

In the theory of Turing machines, the problem investigated by Gödel is presented as a "decision problem". The quintessence of this problem is the conclusion that in any formal system, however extensively and carefully it may have been constructed, there are certain statements that are indeterminate and which cause the Turing machine to become entangled in an endless loop of calculation. The existence of indeterminate statements is in turn an expression of the fundamental incompleteness of such systems. Moreover, subsequent research has shown that—from a mathematical point of view—the non-existence of hidden algorithms cannot be proved. As we will see in Chap. 3, this finding also casts a revealing light on a long und vain debate on the particular nature of living matter.

However, not only in mathematics but also in the empirical sciences one encounters objective boundaries that have important consequences for our scientific understanding of the world. Let us take a look at physics. In physics, too, there are many kinds of boundary. At first sight, these seem superficially to be boundaries set for all natural events, for example by the magnitudes of the fundamental physical constants. The speed of light, setting an upper limit for the speed of anything else, is one of these—as is the absolute zero, or lower limit, of temperature. Another example is the impossibility of constructing a "perpetual motion machine", capable of doing work indefinitely without an energy source: here again we encounter a physical limit.

More difficult to understand are the strange boundaries revealed to us in the microphysical world. There are two main reasons for this: the limitation of the descriptive capacity of human language, and the obscure rôle played by chance in quantum physics.

Let us first examine the linguistic barrier. The concepts dealt with in our language are, in the nature of things, best adapted to describing the "mesocosmic" world of our experience, that is, the world of objects (not too large and not too small) that we can perceive with our sensory organs. In the description of the microcosm, in which the laws of quantum physics apply, the concepts dealt with by our everyday language soon encounter their limits. Here, phenomena can only be described adequately in the language of mathematics. Thus, this limit is actually not a genuine physical boundary, but just an apparent one—one, however, that shows us clearly that human language is only able to depict physical reality to a limited extent.

Alongside the linguistic barrier, another limit confronts us: In the world of elementary physical processes chance plays a central part. This restricts severely the calculability of microphysical events. For example, the point in time when a radioactive atom decays obviously depends entirely upon chance. For this reason the laws of radioactive decay are of a purely statistical nature.

Much thought has been given to the question of whether the world of quantum physics may perhaps, after all, be subject to as yet unknown laws or hidden parameters that determine these elementary processes. This ultimately reduces to the question of whether chance in quantum physics is merely a symptom of our lack of detailed knowledge, or whether it really determines the course of events in an objective sense. At present everything seems to point to the objective nature of the limitations that are placed upon deterministic explanation in quantum physics. A strong indication in favour of this conclusion is a theorem of John Stuart Bell. This theorem, which is supported by experimental evidence, excludes the possibility of "local" theories of hidden parameters. Thus, by all appearances, the possibility that chance makes an essential contribution to quantum phenomena can no longer be ruled out.

The influence of chance naturally places a limit upon the calculability of physical processes. Conversely, though, limited calculability need not necessarily mean that the phenomena under discussion are dominated by chance. On the contrary: unpredictability can even be a direct consequence of deterministic laws. In physics, systems with this property are termed "chaotic" and the phenomenon itself as "deterministic chaos".

The idea that systems may be totally subject to deterministic laws and at the same time behave as though they were governed by pure chance seems to be self-contradictory at first glance. However, this peculiar phenomenon can be made comprehensible by a closer look at the nature of causal determination.

First of all, we need to distinguish between the "causal principle" and a "causal law". The causal principle simply states that nothing takes place without a cause; thus, every event has its own cause. A causal law, on the other hand, is a statement of how cause and effect are linked to one another. This link can be "linear", so that similar causes always have similar effects. However, it can also be "non-linear", so that similar causes can have completely different effects (this issue is discussed in detail in Sect. 8.2).

In the linear case, the system's dynamics can be calculated over a long period, because small indeterminacies in the initial conditions—that is, in the causes—do not change significantly the development of the system. In the non-linear case, on the other hand, even small fluctuations in the starting conditions can have tremendous consequences for the further development of the system because these disturbances, in the course of time, reinforce each other and lead to an avalanche of change. To predict the dynamics of such systems over the long term, the starting conditions must be known with arbitrarily high accuracy. As such accuracy fundamentally cannot be attained, the calculability of systems of this kind lies within very narrow limits.

It is therefore a consequence of the non-linear coupling of causes and effects in chaotic systems that leads to the extreme sensitivity over against small changes in their starting conditions and that makes the dynamics of these systems in practice unpredictable.

Yet the example of deterministic chaos already shows us that our insights into the limits of predictability can themselves be a source of further knowledge. Thus, the physics of chaotic systems has led not only to a deeper understanding of causality and physical order, but also to new insights into the nature of unique processes, such as are characteristic, for example, of historical events (Chap. 8). In this way, the incomputable ultimately appears computable after all.

Deep insights into the structures of the world also arise from the investigation of "floating" boundaries. Such borders are only ostensible, because in their case no exact demarcation, such as might arise from the ontological character of the objects of reality, can be drawn. As an example, let us take an object which we regard as living. Normally, we have no difficulty in recognising a living being as such, because we learned as children to tell the difference between living and non-living entities. Even if small children might sometimes make a mistake when confronted with a cleverly made toy, which looks exactly like an animal, it nonetheless seems as though the ability to distinguish "living" from "non-living" is part of the basic equipment of our cognitive apparatus, and thus is one of the fundamental forms of our implicit knowledge.

The situation, however, is quite different when we come to consider the conceptual delimitation of living beings. This is seen especially clearly in the exact sciences. Any attempt to set up a scientific theory of the origin of life must obviously be able to explain how living matter emerged from non-living matter. This in turn implies a smooth transition between the non-living and the living. Only on this assumption is it possible to sketch out a coherent physico-chemical theory that includes all steps of material self-organisation, from a simple molecule to a living cell. If, however, the transition from non-living to living matter is continuous (or quasi-continuous), then a complete answer to the question "What is life?"—including all necessary *and* sufficient criteria—is impossible, for logical reasons. This in turn means that any scientific theory purporting to offer a complete explanation of the origin of life must inevitably rest upon an incomplete concept of what life is. In this case a certain element of arbitrariness is present concerning the degree of complexity above which a material system is to be regarded as living. For that reason, one can engage in lively discussions about whether viruses are alive or not (see Sect. 3.1).

In the social and political debate about the use of modern biological and medical technologies, the problem of definition is reflected with especial clarity in the problem of defining at which stage of development one should regard a fertilised ovum as a living organism and thus define the beginning of human life. From this enduring debate, however, we cannot draw the conclusion that all attempts to decide this debate by appeal to scientific insights are necessarily doomed to failure, because science is obviously not in a position to state unambiguously the difference between living and non-living matter. Rather, the exact sciences always have the task of abstracting from complex reality. In the present case, the task is to draw a "scientifically" meaningful demarcation line between living and non-living matter, even if in reality no such strict demarcation exists.

Such demarcation lines in science usually emerge from the context of the scientific question being asked or studied. In the case at issue, the specific questions being addressed result in certain properties of "living" matter being emphasised and others being excluded from consideration; only in this way does the scientific concept of "life" constitute an object of research.

In physics, for example, the transition from "non-living" to "living" may be primarily regarded in connection with the origin of certain ordered states of matter, those that are self-preserving. An understanding of this properly requires appeal to thermodynamic principles. Therefore, the physical definition of a living system will involve not only the property of self-preservation, but also (especially) the turnover of free energy. The latter is gained by the decomposition of energy-rich into energy-deficient material, that is, by the process we call metabolism. This process is indispensable, because the living system can only sustain its complex, material order if it receives a perpetual supply of free energy. Only in this way can the continuous production of entropy in the system be compensated for and thereby the "entropic death" of thermodynamic equilibrium be avoided.

In molecular biology, however, other properties of living matter come to the fore. One of these is the fact that all basic processes of life are driven by information. A working definition that takes account of this aspect must therefore give a central place to the concept of genetic information. We could continue to extend the list of different viewpoints from which living matter can be considered, but we would never arrive at a complete definition that could provide us with a logically clear concept for distinguishing it from non-lining matter. All of the characteristic properties that we have mentioned—metabolism, self-reproduction, information and the like—can be found, in some form or other, also in systems that we would clearly denote as non-living.

Now, one might argue that the difficulty in finding a comprehensive concept of what is actually the living is the result of wrong thinking. Perhaps we have been setting out from a false premise by assuming that the transition from the non-living to the living is a smooth one. This immediately prompts an alternative consideration. Maybe the transition is actually a discontinuous one. In that case, it should certainly be possible to draw a sharp borderline between living and non-living matter, and that should make it possible to state necessary *and* sufficient criteria for defining life. However, this would come at an unacceptably high cost. For, a complete definition would have to contain at least one criterion that expresses the ontological difference between living and non-living matter. According to our point of departure, this could only be a *specific* characteristic of life, so that a complete definition of life would always contain a tautological element. Comparable considerations apply to the relationship between mind and matter. Here too, a sharp demarcation cannot be made, because (as we may reasonably assume) the mental properties of matter have developed continuously from its material properties in the course of the evolution of life.

The question "What is life?" finally leads to a unique, and even paradoxical, situation: If we assume that living and non-living matter are in their essence different, then the concept of life can no longer be defined in a logically flawless manner, and the origin of life becomes inexplicable in physical terms. If, on the contrary, we assume that there is no sharp border between living and non-living matter, then a complete physico-chemical explanation is in principle possible—but only at the cost of a reduced concept of life.

The unavoidable incompleteness of the physical concept of life is sometimes presented as a weakness of physicalism. However, this ignores the fact that the exact sciences always operate by using abstractions, simplifications and idealisations. "Point masses", "frictionless movement", "elastic collision", "isolated system" and "reversible process" are typical examples of the numerous idealisations that we encounter in physics. The physical concept of life is no exception.

2.3   The Whole and Its Parts

To emphasize it once more: Abstraction, simplification and idealisation constitute the methodological foundations of the exact sciences. They determine the fundamental nature of scientific knowledge. For precisely that reason, the exact sciences cannot embrace the richness of the reality of our life, in all its breadth and depth.

In the public perception of science, this has always been a source of criticism, as it is often believed that even the methodological requirements of the exact sciences restrict fundamentally the knowledge that they convey, or cast doubt upon it. However, this is a grave misunderstanding: the methodological restrictions of science do not place any limits on its discoveries as such, but rather, at worst, on the scope of its discoveries.

The exact sciences, the tool of which is the analytical dissection of a scientific problem, are only able to explain narrow aspects of the phenomena that are under investigation. Accordingly, our scientific understanding of the world consists of a mosaic of innumerable components of knowledge, of which each component can be critically assessed and, if necessary, replaced by a new one.

The general procedure followed by the exact sciences is to explain the whole on the basis of the combined effect of its parts. In public debate this is often branded as the "mechanistic" world-view of the exact sciences and, it is claimed, as totally inadequate to explain the complex reality of our life or, especially, the idiosyncrasies of living Nature. Instead, critics of the exact sciences repeatedly demand a "holistic" understanding of reality. This criticism is irritating insofar as it not only paints a completely wrong picture of the sciences, but also instrumentalises this picture to feed a socio-political programme that is inimical to both science and technology. It usually culminates in the accusation that the scientific method, above all in the natural sciences, leads to a wrong understanding of reality and thus promotes ruthless exploitation of Nature by modern technologies. Some social philosophers do not even hesitate to dub science and technology as an "ideology", the intention of which is entirely to obtain dominance over mankind and Nature (see [5]).

Mistrust of the exact sciences lies deep. For many people, progress in science and technology is regarded as in reality retrogressive and as leading to a cold and unfeeling society, one that threatens mankind and Nature alike. Needless, to say, the depressing prospects that these critics of science and technology project onto the screen cause fear, which in turn engenders a yearning to return to an unspoiled state of Nature. It is claimed that only an organic understanding of Nature, one that embraces the whole, can match up to the problems caused by the scientific and technological handling of Nature and the permanent, growing ecological problems. Little wonder, therefore, that many people are trying to replace the spectre of a reality regulated by mechanical laws alone with an old-fashioned and romantic vision of Nature.

The accusations directed against the exact sciences are always the same. First, it is claimed that the "reductionistic" programme—directed as it is by dissection, simplification, abstraction and idealisation—cannot but lose the "whole" perspective. Secondly, it is claimed that the causal-analytical method which endeavours to explain "the whole" by appeal to the combined behaviour of its parts is in itself completely unable to conceive of the fundamental nature of life, simply because any living system is an irreducible whole, a self-contained circuit of causes and effects.

The consequent demand for a holistic access to reality has in the meantime become a mantra of the "alternative" science scene. What we are supposed to understand by this, however, remains a closed book. It is perfectly correct to characterise the essence of an organic whole as a cyclically linked system of causes and effects. However, a real understanding of the whole—insofar as "understanding" is to be taken as implying a causal explanation—requires the analytical dissection of the whole. Otherwise the understanding of the whole is merely surrogated by a superficial overview of the whole.

The arguments adduced to cast doubt upon the reductionistic method are extremely vague. By appeal to the autonomy of wholes, it is demanded (usually in highly rhetorical terms) that analytical thought which is associated with mechanistic models of the reality should be augmented with "holistic" or "connected" thinking. Admittedly, one wonders which science is being criticised here. Modern science has long outgrown its infant belief in a simple mechanistic world. In fact, if we wish to attribute any real meaning to the holistic (but trivial) thesis according to which "the whole is more than the sum of its parts", then there is no better place to look than modern physics. Quantum physics has even developed a proper terminology in order to take account of the holistic phenomenon of "entanglement" of quantum objects. Likewise, the holistic thesis of macrodetermination, according to which the whole determines the behaviour of the parts, is by no means foreign to the exact sciences. To see this, one does not need to look further than the "law of mass-action" in chemistry. When a chemical system is in a state of equilibrium, the individual molecules in the system are unaware of this. However, the system as a whole determines, by a mechanism of negative feedback of fluctuations, the behaviour of its molecules: the greater the (random) deviation from the equilibrium state, the stronger is the tendency of the system to move back toward this state. It is precisely this that is expressed by the term "mass action" (see Sect. 7.2).

As already pointed out by the biologist Peter B. Medawar, in reality the "reductive analytical-summative mechanist" does not exist. Rather, "he is a sort of lay devil invented by the feebler nature-philosophers to give themselves an opportunity to enjoy the rites of exorcism" [12, p. 144]. The criticism of the analytical sciences is in fact a relic of the romantic epoch. Then as now, the emphatic rejection of the mechanistic worldview was ultimately directed against the forward march of technical mastery over Nature. And, at the same time, loud demands were made that the "blind" and "unimaginative", mechanistic approach to research should be confronted by a holistic, i.e., organismic understanding of Nature (see Sect. 1.4).

Since those days, the spectre of mechanism has again and again been evoked, a purported ideology in science and technology, which—in the thrall of a mechanistic myth of progress—are set to destroy the natural basis of human life, without taking any account of the holistic character of the animate world. Yet experience has shown that in actual fact it is not analytical, but rather holistic, thinking that brings danger. If the whole determines the behaviour of its parts, or—as claimed by the representatives of the so-called philosophical holism—if "the higher entity is always the prime mover of the lower" [13, p. 356], then the temptation is very close at hand, especially in totalitarian political systems, to apply this thought in legitimising authoritarian structures of power in the society. Anne Harrington, who has made a profound study of the history of holistic thinking in the first half of the 20th century, has shown that holistic ideas have indeed made a substantial contribution to political indoctrination and to the propagation of totalitarian ideas [7].

2.4 Concerning Hypothetical Truth

There are also boundaries that are imposed upon science by society. These boundaries add up to a complex mesh of prohibitions and regulations that lay down norms for the process of research. These normative boundaries serve the purpose of controlling the gain of scientific knowledge and of steering technological progress in a "forward-looking" direction. But where should science and technology go? On what can the norms that we impose upon science possibly be based? Should we refer to norms that are based upon metaphysical grounds of being and lay claim to absolute truth, or should we only refer to such norms as can be derived from scientifically based knowledge itself that has basically an empirical, and therefore always a preliminary, character?

These and similar questions come to play a decisive role in situations where appeal is made to the "responsibility" of science. The exercise of responsibility by restricting the gain of scientific knowledge does not only mean that we have identified the problems that are posed by scientific and technological progress; it further implies that the alleged problems are in fact "genuine" problems. For this reason we cannot separate the concept of norm from that of truth. However, as the history of philosophical thinking shows, the concept of truth is many-layered. It includes absolute, empirical, hypothetical, contingent, historical, logical, mathematical, objective, ontological, subjective, practical, theoretical and utilitarian truths, and many more besides. Furthermore, all conceptions of truth demand, in some form or other, a theory of truth. The most prominent of those theories are the correspondence theory, the consensus theory, the discourse theory and the coherence theory.

More important, though, is the fact that the problem of fluid boundaries, discussed above, arises also with regard to the question of "truth". In this case the problem is expressed in the phenomenon of unsharp truth-values. This phenomenon, which was already known to the ancient Greeks, has been passed down to us in the well-known trick question of Eubulides: How many grains of sand make up a pile? One grain of sand is certainly not a pile. Neither are two. Obviously, increasing the number of sand grains by one does not bring about the critical difference between a pile and a non-pile. Nevertheless, a sufficient number of grains of sand quite definitely constitutes a pile.

There are numerous examples in which the boundaries between "true" and "false" are floating ones, so that the truth content of statements may decrease on a slippery slope of diminution or vice versa. In basic research a special logic for this has even been designed, the so-called "fuzzy logic", which takes account of the existence of floating truth-values. In quantum physics, too, there have been attempts to encounter epistemological difficulties arising there by creating a special logic in which, for example, probabilities are assigned to the truth-value of statements. To avoid disruption of the logical foundations of our thought, repeated efforts have been made to justify the unity of logic and to trace all logical forms back to the binary logic with which we are familiar and in which only true and false statements are recognised.

In view of the inflationary numbers of concepts of truth one may be excused a certain helpless perplexity, asking, along with Friedrich Nietzsche, what human truth ultimately is. Is it, as Nietzsche mocked, nothing other than the "irrefutable errors" of mankind [15, p. 265]? When dealing with purported truth, are we in reality dealing merely with truths devoid of truth content? Is truth ultimately only the difference between errors?

Even if, with these questions, we are up against the limits of human thought— we still cannot relinquish the concept of truth, as any thought or action that abandons in advance all claims to truth naturally loses touch with reality. In fact, any attempt to pursue the idea of truth *ad absurdum* can only be taken seriously if it at least claims to be true in itself—and in so doing tacitly admits the existence of what it opposes, namely, the idea of truth. This self-referentiality of truth cannot be disrupted by any critical theory of truth. Conversely, no theory of truth can ever prove the existence of absolute truths. Such efforts have no hope of success, if only because there is no Archimedean point outside of truth that would allow us an absolutely trustworthy perspective view of the truth issue. This is the actual essence of the aporia that is hidden in the concept of truth.

In consequence of this, the debate on the question of truth has never ceased to run in circles, although it admittedly has revealed, in the course of its long history, numerous truths about truth itself. All attempts to unify the concepts of truth or to deduce one such concept from another are inevitably condemned to failure, because it is precisely the multiplicity of concepts of truth that reveals a sophisticated and differentiated understanding of reality. On the other hand, an essential property of truth is that it cannot be forced into line. What remains is the idea of truth as a "regulative" idea, one that appears indispensable, because without the idea of truth the human endeavour to acquire reliable knowledge of the world would at once lose both its goal and its content.

In keeping with this, scientific knowledge of the world can also never claim more than hypothetical truth. This insight was stated clearly by the philosopher of science Karl Raimund Popper, who made it the basis of his treatise "Logik der Forschung", which appeared in 1934. Today it is generally accepted that the empirical sciences can never provide absolute certainty about the validity of any discovery. Rather, all scientific knowledge is merely of a provisional and hypothetical nature. Yet there remains, for the empirical sciences, the instrument of observation and experimental test, by which a claim to truth can at any time be critically examined and, perhaps, be refuted. It is in this that a scientific understanding of reality shows its strength over against any dogmatic world view that purports to possess absolute truth.

This automatically answers the initial question of which norms we should refer to in our dealings with science and technology. These can only be norms that in turn rest upon well-founded scientific knowledge. For only on the basis of scientific knowledge can a critical and forward-looking consciousness develop, one that is capable of taking issue in a rational manner with scientific and technological progress. This in turn presupposes the unrestricted freedom of gaining scientific knowledge.

2.5   We Must Search, We Will Search

The exact sciences can only live up to their own criteria of critical and enlightenment-directed thinking by drawing a sharp line between themselves and all kinds of pseudoscience. According to Popper, this demarcation is made possible by the criterion of "falsification": a hypothesis or a theory can only claim scientific validity when it leads to experimentally testable predictions and thus, in principle, is falsifiable (refutable).

Indeed, most theories in physics meet the demarcation criterion demanded by Popper. But what about the scientific theories that purport to describe the history of Nature? Historical processes, however, appear to be unpredictable in principle. A favourite example of this is Darwin's theory of evolution; this theory admittedly allows causal explanations, but makes practically no predictions. The most important reason for this is that the paths of evolution depend to a high degree upon chance. The issue is exacerbated by the enormous material complexity of living beings, even at the lowest stages of development, so that there is only very limited room for precise calculation. For this reason Darwinian theory, compared with physical theories, has only very limited predictive power. It has even been asserted that the Darwinian principle of "survival of the fittest" is a mere tautology, because the central concept of fitness is determined solely by the fact of having survived. In view of this, Popper went as far as to dismiss evolution theory as a "metaphysical" research programme, one which does not measure up to the status of a scientific theory—a viewpoint that he admittedly later revised [16].

Beside the general question regarding the scientific status of Darwin's theory there are other problems that challenge the idea of Darwinian evolution. One is known as the statistical problem of the origin of life. This encapsulates the fact that the spontaneous formation of a living being by the random association of its material building-blocks is vanishingly small (see Sect. 3.2). Further, the statistical analysis has shown that not even a simple biomolecule, carrying some biological function, could originate by pure chance during the past lifetime of our universe.

The statistical problem seems to indicate that the existence of life must be regarded as an enigma of Nature. Under these circumstances it is all the more astounding that one of the leading molecular biologists of the 20th century, Jacques Monod [14], invoked precisely the statistical objection in order to found his "chance hypothesis" of the origin of life. How is this to be understood? Here one has to remember that probabilities say nothing about the actual occurrence of a single event, but only about its relative frequency. For this reason it is perfectly possible to attribute the existence of life on Earth to a singular chance event, even though the probability of this happening is practically zero. The low probability only indicates that this event is not *reproducible* within the limits of the universe. In this sense Monod believed that he could interpret the origin of life as having been a once-off event in the entire cosmos, one that with a probability bordering upon certainty would never be repeated.

Even if we cannot exclude the possibility that life originated as a singular, chance event, a hypothesis of this kind is completely unsatisfactory for the scientist. It can hardly be the goal of scientific research to attribute the phenomena demanding explanation to the effect of singular chance. It is rather the task of science to understand the phenomena in terms of the law-like behaviour of Nature. Scientists remain equally unconvinced by arguments from the extreme opposite camp which assume hidden laws of life, or a cosmic or a divine plan that guide the processes of Nature. Such hypotheses are by their very nature non-falsifiable, because the non-existence of inscrutable plans, final causes, life forces and the like can never be proved [9]. Conversely, no-one has yet succeeded in adducing the most rudimentary evidence for the presence of life-specific principles in Nature.

As was expected, the open questions regarding the origin of life have become a playing-field for all conceivable kinds of pseudoscientific theory, of which the most influential today, camouflaged as a religious movement, styles itself "intelligent design". However, we cannot conclude from the present gaps of scientific explanation that there is basically no solution within the framework of science. In fact, the history of science shows that science has steadily closed gaps in explanation that first seemed to have no solution.

The same applies to the statistical problem of the origin of life. In place of the wrongly set-up decision "chance" *or* "necessity", modern biology has long offered a satisfactory reply to this question, one that joins up chance *and* necessity in the Darwinian sense (Chap. 3). The solution of the statistical problem has become possible through the discovery that natural selection is a universal principle of Nature—one which under the right physical conditions also operates in the realm of molecules, where these conditions can steer the selection of those molecular structures that are able to organize themselves into a precursor of the living cell [4]. On the one hand, this process is subjected to chance, while on the other the results of chance are "evaluated" by the law-like action of natural selection, so that the overall process of the origin of life, as denied by Monod's hypothesis, can *in principle* be a repeatable process. Moreover, at the molecular level of evolution precise predictions have become possible, which can even be tested experimentally.

With these achievements, the accusation of tautology is refuted, as is the assertion that evolution theory is merely a metaphysical research programme. In the face of the massive challenges that time and again have been raised against Darwinian evolution theory, we are obliged to acknowledge the triumph of Darwin's idea by the development of the molecular theory of the origin of life. Within the framework of this theory the evolution of biological macromolecules cannot only be justified theoretically; it can also be simulated in the test-tube.

The experimental technique for the study of the self-organization and evolution of biological macromolecules has now been developed to the extent that the experiments can even be performed by automata (Sect. 3.4). Moreover, such "evolution machines" are able to start a process of evolutionary optimization from any area of information space and thus liberate natural evolution from the constraints of its pursued routes. Already today, we can anticipate the time when we will be able to investigate the inexhaustible potential of life in all directions and bring it, supported by new techniques, to its full development. This marks the beginning of the dissolution of yet another fundamental boundary of our understanding of reality—the boundary between "natural" and "artificial" evolution.

Admittedly, this fascinating technical possibility opens up another range of novel questions; the most obvious one is that of the content of the genetic information that will be generated by artificial evolution. This in turn requires an approach to the semantics of information within the framework of the exact sciences (see Chap. 5). In this respect, promising first steps have recently been taken, ones that also throw a new light on the nucleation of semantic information in prebiotic matter [10].

Looking back on the scientific developments of the last hundred years we can conclude that not much is left of the seven "world enigmas" that Du Bois-Reymond regarded as fundamentally unsolvable. (1) The nature of force and matter has been made comprehensible by twentieth-century physics, in a depth that is without compare in our understanding of the world. (2) The question of the origin of movement, which seems to be a metaphysical relic of the Aristotelian doctrine of movement, has today found a scientific answer in the physics of self-organisation. (3) The problem of the origin of life has also become accessible to science, and its main outline is already well enough understood for it to be declassified as a world enigma. (4) The apparently purposeful, planned, goal-oriented organisation of Nature has found an explanatory basis in the modern theory of biological evolution. (5) Simple sensory perception, in modern brain research, has today come a long way, especially for the processes involved in visual perception. The seven world enigmas have thus shrunk to two: rational thought and the origin of language, and free will. Here we have without doubt arrived at a frontier of present-day research. However, nothing contradicts the expectation that these problems, too, will one day lose the aura of mystery that today continues to surround them.

References

1. Du Bois-Reymond, E.: Über die Grenzen des Naturerkennens. Veit & Co., Leipzig (1872)
2. Du Bois-Reymond, E.: Culturgeschichte und Naturwissenschaft. In: Vorträge über Philosophie und Gesellschaft. Veit & Co., Leipzig (1874)
3. Du Bois-Reymond, E.: Die sieben Welträthsel. Veit & Co., Leipzig (1880)
4. Eigen, M.: Self-oganization of matter and the evolution of biological macromolecules. The Science of Nature (Naturwissenschaften) 58, 465–523 (1971)
5. Habermas, J.: Technology and science as an ideology. In: Toward a Rational Society. Student Protest, Science, and Politics. Beacon Press, Massachusetts (1971). [Original: Technik und Wissenschaft als ‚Ideologie', 1969]
6. Haeckel, E.: The Evolution of Man, vol. 1. The Werner Company, Akron [3](1900). [Original: Anthropogenie oder Entwicklungsgeschichte des Menschen, 1874]
7. Harrington, A.: Reenchanted Science. Princeton University Press, Princeton (1999)
8. Hilbert, D.: Naturerkennen und Logik. The Science of Nature (Naturwissenschaften) 18, 959–963 (1930)
9. Küppers, B.-O.: Information and the Origin of Life. MIT Press, Cambridge/Mass. (1990). [Original: Der Ursprung biologischer Information, 1986]
10. Küppers, B.-O.: The nucleation of semantic information in prebiotic matter. In: Domingo, E., Schuster, P. (eds.): Quasispecies: From Theory to Experimental Systems, pp. 23–42. Springer International Publishing, Switzerland (2016)
11. Lorenz, K.: Kants Lehre vom Apriorischen im Lichte gegenwärtiger Biologie. Blätter für Deutsche Philosophie 15, 94–125 (1941)
12. Medawar, P.B., Medawar, J.S.: Aristotle to Zoos. Harvard University Press, Cambridge/Mass (1983)
13. Meyer-Abich, A.: Naturphilosophie auf neuen Wegen. Hippokrates, Stuttgart (1948)
14. Monod, J.: Chance and Necessity. Vintage, New York (1971). [Original: Le Hasard et la Nécessité, 1970]
15. Nietzsche, F.: The Gay Science. Cambridge University Press, Cambridge (2001). [Original: Die fröhliche Wissenschaft, 1882]
16. Popper, K.R.: Unended Quest. Routledge, London (1992)

Chapter 3
How Could Life Have Originated?

"In the beginning the earth was without form, and void." the Bible says. Was the earth once as hostile to life as the Martian landscape shown here? How did life originate? Did life require a divine plan, or could it simply arise by chance? Do the known laws of physics and chemistry suffice to explain the origin of life?

© Springer International Publishing AG 2018
B.-O. Küppers, *The Computability of the World*, The Frontiers Collection,
https://doi.org/10.1007/978-3-319-67369-1_3

3.1 What Is Life?

The question of the origin of life can be examined from two perspectives. On the one hand we can try to reconstruct, as exactly as possible, the historical prerequisites and conditions that facilitated the formation of life on the early Earth. One the other hand, we can also look for the general principles and laws that govern the transition from the non-living to the living. In this case it is not the historical aspects of early evolution that are of interest, but rather the regularities that underlie this process.

Regarding the historical perspective we can dispense with a definition of "life". Our main task is to describe the historical circumstances by which the known forms of life came into being and developed. However, as the primordial forms of life from which the biosphere emerged no longer exist, we can only try to make reasonable suppositions about the geological and physico-chemical conditions that prevailed on the Earth during the prebiotic phase of evolution. This restriction makes the historiographic approach subject to the uncertainties and the gaps in knowledge that are typical of this kind of analysis: whether life originated in the ocean, in a pond, in hot springs, in porous rocks or against some other prebiotic backdrop will therefore ultimately remain an open question, even though the various accounts put forward will differ in plausibility.

Looking for law-like mechanisms, in contrast, we are dealing primarily with the general laws and principles that gave rise to the nucleation of living matter. Their validity does not depend upon the historical conditions under which life once evolved, since laws and principles, by their very nature, describe processes that are independent of space and time. Ideally, complete knowledge of all the relevant laws and principles should even allow us to investigate the decisive steps towards life under laboratory conditions.

However, in contrast to a mere historical description, the law-like reconstruction of the origin of life inevitably raises the fundamental question of the difference between non-living and living matter. For that reason, any attempt to develop a general theory of the early steps towards life requires a satisfying definition of living matter. However, as we have seen in Sect. 2.2, this task leads us into immense difficulties, because the definition we are seeking depends always upon the particular context of our investigation.

The question "What is life?" has *de facto* no final answer. Rather, it induces a cascade of further questions that reflect the various perspectives under which the problem of life can be approached in science. A physicist, for example, would at first point out that a characteristic property of life is its extremely complex state of material order. This immediately raises the next set of questions: "How can we specify this order?" and "What kind of complexity do we encounter in living matter?". In fact, what do the terms "order" and "complexity" mean when they are applied to biological systems? These questions would probably attract the attention of a molecular biologist, who would reply: The complex order of life is a functional order that is governed by information. However, that immediately provokes further

questions: "What is information?, "Can information be a property of matter?",
"How could semantic, i.e., meaningful information, as is present in the genome of a
living being, originate?". At that point, doubtless an evolutionary biologist would
intervene and say something to the effect that semantic information originated
during the early phase of evolution.

However, the questions have now become: "Is the ability to evolve a general
property of matter, which was already present in prebiotic matter?", "Can those
aspects of genetic information that refer to its sense and meaning ever become the
subject-matter of the exact sciences?". This is the point at which we might turn to a
philosopher—but there are serious doubts as to whether he, too, might not run into
difficulties with these questions. The reasons for this are clear enough: "concepts in
science must first be brought into focus by more comprehensive knowledge or,
ultimately, by a theory" [25, p. xii]. In other words: basic terms of science—such as
order, complexity, information, semantics etc.—need to be undergirded by a theory
before they can be defined crisply. Conversely, however, every scientific theory is
based upon fundamental terms that make up the foundations of the theory.

Thus, the formation of terms and the formation of theories are indissolubly
interrelated, and the explanatory generality of a theory increases with the sharpness
of its terms. The considerations that follow are intended to make clear how this
reciprocal relationship applies to the issue of the origin of life. Moreover, we shall
see how the analysis of this interrelationship leads to novel insights into the possibilities and the limitations of objective knowledge in biology.

To guide us, we shall frequently refer to the compendium of questions that we have
already raised in Sect. 2.2. Let us start once again with the question "What is life?"
Comparing a range of living organisms shows us two properties that they all possess:
they have a metabolism and they are capable of self-reproduction. As a third property
one might add mutability over generations—that is, inaccuracy in self-reproduction.

Metabolism, self-reproduction and mutability seem to be suitable criteria to
distinguish between living and non-living matter. Yet these three distinguishing
characteristics are only *necessary* criteria, but not *sufficient* ones: they still spread
the net too wide. Consider, for example, the viruses. They are among the simplest
organisms that we know. However, they only fulfil the criterion of metabolism
within their host cell, as they do not possess an independent metabolism. Outside
their host they behave as inert, non-living structures that can even be crystallized.

A virus that has been thoroughly studied in this respect is the so-called tobacco
mosaic virus, which spreads in the leaves of tobacco plants. In the electron
microscope the characteristic rod-shaped outline of this virus is seen clearly
(Fig. 3.1a). Analysis of these particles has shown that they consist basically of two
classes of biological macromolecules: a long, spirally wound thread of nucleic acid
(RNA), on which more than 2000 identical proteins are arranged in rows
(Fig. 3.1b). The protein units make up a stable protective outer coating for the viral
RNA, which contains the hereditary information.

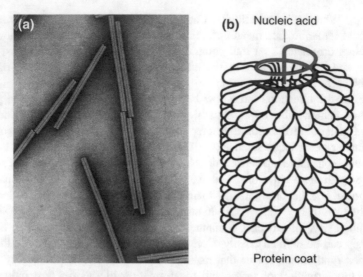

Fig. 3.1 The tobacco mosaic virus as an example of the shifting boundaries between living and non-living matter. **a** Electron-microscope photograph of tobacco mosaic virus at a magnification of 300,000 (from [2]). **b** Schematic illustration of the structure of the tobacco mosaic virus. Each virus particle consists of a nucleic acid, which carries the genetic information, and 2130 identical protein units arranged in a spiral around the nucleic acid. The protein units form a stable protective coat around the genetic material. As soon as the virus has infected a cell, the protein coat is sloughed off and the genetic material is released. The cell's metabolism is then re–programmed to produce new virus particles

The macromolecular components of the virus are held together by weak physico-chemical interactions: hydrogen bonds and Van der Waals forces. In biology, weak interactions always play an important part in situations where large molecular structures are continually built up and broken down. The stability of such structures, like that of a zip, is provided by the mutual, co-operative stabilisation between the individual bonding interactions; however, the individual bonds are weak enough to be disrupted rapidly if this is needed.

The viruses are themselves subject to perpetual assembly and dismantling. For example, the tobacco mosaic virus, after penetrating and entering its host cell, sloughs off its protein coat, thus ensuring that its genetic material is released and is able to take part in the host cell's metabolism. The subsequent assembly of new virus particles takes place through the spontaneous adhesion between the nucleic acid and the protein units. As complex as this process is, it can be replicated in the test tube and its individual steps can be studied. Dissociation and reconstitution experiments of this type were first performed successfully in the mid-1950s [12]. The experiments showed that one can indeed dissect tobacco mosaic viruses into their molecular constituents and then put the parts together again to produce infectious viral particles. Even though the primal identity of each virus is lost in this process, the reconstituted viruses are completely indistinguishable from natural viruses. Here, the everyday understanding of "living" and "non-living" loses its meaning.

In the border zone between living and non-living matter we run directly into a problem that in the philosophy of science is referred to as the problem of "reduction". If we set out from the working hypothesis that all the processes of life are, in principle, reducible to the laws of physics and chemistry, then this presupposes that the transition from "living" or "non-living" is a gradual one. This in turn means that it is impossible to find necessary *and* sufficient criteria to define the phenomenon "life" (see Sect. 2.2). Therefore any such definition must inevitably remain incomplete.

However, we can also look at this problem from a completely different perspective. We can relinquish the goal of achieving a complete definition of life and simply ask what physical principles underlie the elementary characteristics of life such as metabolism, self-reproduction and mutability.

Let us first consider metabolism. Physically, this phenomenon is fairly easy to understand. Metabolism is nothing other than the turnover of "free" energy. This is the name given to the form of energy that is capable of doing work. Free energy is needed by living systems to sustain their complex material states. Without it, they would decay into the "dead" state of thermodynamic equilibrium, that is, the state of maximum entropy and the greatest possible material disorder (see Sect. 7.2).

The reproductive self-preservation of living matter appeared for a long time to evade a physical interpretation. Up to the 1930s, this phenomenon was regarded by physicists as so mysterious that Niels Bohr, a co-founder of modern physics, seriously advocated the idea of the irreducibility of living beings [1]. The decisive breakthrough concerning this problem had to wait until the epoch-making discovery of the molecular mechanism of heredity by Francis Crick and James Watson in 1953. This made clear at a stroke that the reproductive self-preservation of living systems is a direct consequence of the chemical properties of the genetic material.

The molecules of heredity belong to the class of nucleic acids (Fig. 3.2). These are long, thread-like molecules, built up from four kinds of basic building-blocks that are referred to as "nucleotides". In the nucleic acid molecule, the nucleotides are arranged in a row like beads on a necklace. In this form, like the letters in a piece of writing, they encode the entire information for the construction of an organism. The following sequence

...UGCACGUUCUCCAACGGUGCUCCUAUGGGG...

represents a typical example of the linear structure of genetic information. It is a small excerpt from the genetic blueprint of the bacteriophage MS2, a virus that infects bacteria [11]. If a nucleotide is replaced by one of the other three nucleotides, then this mutation can, like a misprint in a book, distort the meaning of the genetic text and, in the worst instance, may lead to the complete disruption of the functional order, and thereby to the death, of the organism.

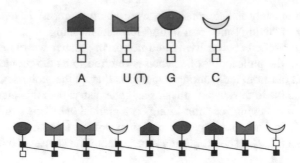

Fig. 3.2 Illustration of the way in which a nucleic acid is built up. The nucleic acids are chain-like molecules and comprise four basic building blocks ("nucleotides"). Each nucleotide in turn is composed of three small molecules: an organic base, a sugar molecule and a molecule of phosphoric acid. The sugar comes in two forms, ribose and deoxyribose; accordingly, there are two kinds of nucleic acid, ribonucleic acid (RNA) and deoxyribonucleic acid (DNA). A further minor difference between the two is that RNA uses the base uracil and DNA the base thymine. The bases are usually referred to by the initial letters of their chemical names: A(denosine), G(uanine), C(ytosine) and U(racil)/T(hymine)

The molecular structure of nucleic acids provides a simple physical mechanism for their self-reproduction. This mechanism is based primarily upon the principle of complementary pairing of nucleotides. In fact, the four nucleotides can be grouped into two "pairs", whereby each member of a pair can "recognise" the other member, and join up with it specifically (Fig. 3.3). Thus, the nucleotides G and C form a stable pair, as do A and T (or A and U). This property of the nucleotides provides the basis for the identical replication of a nucleic acid molecule. By joining up with complementary nucleotides, the nucleic acid to be copied first generates a "negative" copy and in a second replication step this is converted back into a "positive" one (Fig. 3.4). In the chromosomes, the two complementary DNA strands are joined up into a double helix resembling a screw (Fig. 3.5). In reproduction they become separated, like the halves of a zip, and each is then joined up to a newly formed complementary strand.

The genetic information of a living organism is thus encoded in the unique sequence of the nucleotides of its genome. If the genome is transferred to a suitable

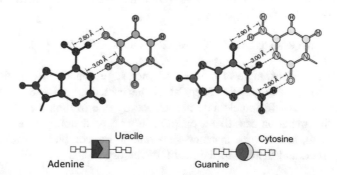

Fig. 3.3 Complementary base pairing. By way of hydrogen bonds, the bases A and U (or T) and the bases G and C join up to form base pairs that are geometrically almost identical

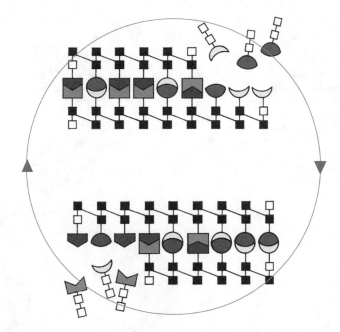

Fig. 3.4 Cross-catalytic reproduction of a nucleic acid

medium, for example that of an egg cell (ovum), then the information encoded in the genome is expressed in material form. This process takes place, as far as we can see today, completely within the framework of the known laws of physics and chemistry.

Modern methods for determining the sequences of nucleic acids have revealed to us the amazing complexity of the information that is encoded in the genome of a particular organism. It has become clear that the construction and reproductive self-preservation of a living system requires an incredibly complex, convoluted hierarchy of instructions. This at the same time means that genetic information cannot simply be read off, step by step, and immediately translated into the material structure of the organism. Rather, there has to be a permanent feedback loop connecting the genome and the gene products, so that the organism can develop out of the genome in a stable manner—that is, with preservation of its internal statics—in successive steps. The overall process is thus a permanent reshaping of the structure that the organism has at any given moment of its development. To guarantee the coherence of this process, the exchange of information inevitably takes on the form of a permanent communication process between the whole and its parts.

The idea of information is undoubtedly a key concept in understanding living matter in general as well as for the origin of life in particular [17]. From this perspective, we can also offer an answer to the central question "What is life?":

$$Life = Matter + Information$$

Fig. 3.5 The molecular
mechanism of heredity.
The DNA, as carrier of
hereditary information,
resides in the chromosome as
two complementary strands,
twisted round one another in a
double "screw thread" (helix).
When the DNA is reproduced,
the two strands separate from
one another and a new
complementary strand forms
on each of the separated
strands

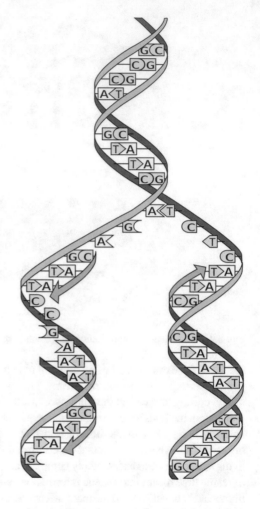

That information is indeed a natural entity will become clearer in Chaps. 4 and 8.
Here it will suffice to note that the concept of information in biology can be given a
strict physical foundation. In fact, genetic information turns out to be a specific type
of physical boundary imposed upon matter, one that places constraints upon the
action of natural laws in the living organism.

    Let us enlarge a little on this idea. Natural laws are usually described in the form
of differential equations. These can only be solved unambiguously if the initial
conditions, under which the solutions are sought, can be stated. However, in certain
circumstances, feedback between the system's dynamics and its initial conditions
can take place that modifies the initial conditions. In systems of this kind the initial
point of the dynamics ultimately becomes lost in the course of the systems's

development. The initial conditions then adopt the character of "boundary conditions" or "constraints", which perpetually channel the course of the system's development. In such a case one speaks of a "self-organising" system.

If we bear in mind that all the processes of life are governed by information, then it becomes easier to understand why theoretical biology today displays many of the traits of the information sciences—with the primary goal of shedding light upon the structure of genetic information and its functional mechanism. In this respect, biology once seemed to have run into an insurmountable barrier. Only a few decades ago, the determination of a relatively short sequence of genetic information (say 100 nucleotides) represented some two years of hard laboratory work. Yet even the simplest viruses have genomes amounting in length to several thousand nucleotides. To read off the genome of a virus would, by the standards of those days, have taken a whole century of toil by molecular biologists. However, even in this case science has made possible the seemingly impossible, and has overcome an obstacle to experimental research that seemed insurmountable. Within only a few decades, it proved possible to accelerate and to automate the procedures for sequence determination to such an extent that today huge sections of a gene can be effortlessly sequenced in a remarkably short time. Thanks to this progress, we now have a complete working sequence of the human genome.

3.2 How Complex Is Living Matter?

Before going on to pursue the question of how genetic information may have originated, let us first try to obtain a more detailed idea of the complexity of living organisms. A first rough measure of the complexity of a living system is the quantity of information that is contained in the molecules of heredity. The genomes of simple organisms, such as the RNA viruses, generally comprise several thousand nucleotides. However, viruses do not have an independent metabolism; the smallest organisms that are autonomous in this respect are the bacteria. Their genetic information—for example, that of the bacterium *Escherichia coli*—runs to several million nucleotides. Yet the human genome contains several thousand million nucleotides!

Nonetheless, the mere "length" of a genome, expressed in the number of its nucleotides, still fails to give real insight into its material complexity. For example, bacteria and humans differ in genome length by a factor of only about 1000. Yet this comes nowhere near reflecting the real difference in their complexity. To make the difference clearer, we need to assess the various levels of complexity of organisms by looking at the range of variation with which their information can be encoded.

The variation is calculated from the number of possible ways in which the nucleotides in a nucleic acid of a given chain length can be arranged. It is a direct measure of the structural richness that may become the basis of the evolution of functional complexity in living matter. Ultimately this measure characterizes the degree of evolutionary uniqueness of the particular organism.

Starting from a given nucleotide sequence the number of alternative sequences that are combinatorially possible can easily be calculated: For a nucleotide sequence consisting of n building blocks, each position in the sequence has four ways in which it can be occupied by one of the four basic building blocks; the total number of possible arrangements is thus:

$$4 \times 4 \times 4 \times \ldots \times 4 = 4^n \approx 10^{0.6n}$$

Let us now look at some real numbers. The genetic programme of a virus comprises as a rule between one thousand and ten thousand nucleotides. According to this, the genome of a virus—the lowest level of functional complexity – has already 10^{600} alternative sequences. This is a vast number. For higher organisms, the complexity continues to rise dramatically: the genome of the bacterium *Escherichia coli* has $10^{2\ 400\ 000}$ and that of human beings more than $10^{600\ 000\ 000}$ combinatorial alternatives.

Numbers of this magnitude cannot justly be termed merely "astronomical". They belittle any attempt to imagine them. This makes it clear at a stroke that no event of pure chance, no random abiogenesis, could ever be in a position to generate the genome of even the simplest organism. Even if we take account of the virtually boundless extent of the universe, this argument remains unaffected.

In fact, there is no doubt that the dynamic complexity of the universe could never extend to making a random origin of life even remotely probable. This becomes evident when one calculates the number of all possible physical processes that have taken place since the universe began. At first sight it may seem impossible to determine this number. However, for our purpose we do not need to calculate the actual complexity of the universe; we simply need to estimate its *upper limit*, and this can be done quite simply.

According to all we know at present, the total mass of the universe, expressed as a multiple of the mass of a single hydrogen atom, is about 10^{80}. The age of the universe has been estimated to be around $13{,}6 \times 10^9$ years, corresponding to some 10^{17} s. We can also express the age of the world in the smallest physical unit of time, the so-called "Planck time", which is about 10^{-43} s; below this time frame quantum-gravitational effects lead to fault lines in space-time, and the event character of the world becomes blurred. The age of the universe in Planck-time units is approximately 10^{60}.

With the total mass and the age of the universe we have two basic values that enable us to make a rough estimate of the upper limit to the dynamic complexity of the material world. This limit would have been reached if, since the beginning of the universe, every material particle had interacted again and again with other particles within the shortest physical timeframe. The greatest number of possible processes is thus the product of the numbers 10^{80} and 10^{60}. This means that since the beginning of the universe not more than 10^{140} elementary processes could possibly have taken place. Even if this number is still approximate, and may one day need to be corrected by several orders of magnitude, it still shows clearly that, with a probability bordering upon certainty, life could never have arisen in our universe by a singular chance event.

A genome with a prespecified nucleotide sequence cannot be generated by playing molecular roulette, because even in the lifespan of the whole universe only a tiny fraction of the total number of combinatorially possible sequence alternatives could ever have been tried out in a random synthesis. Therefore, even at the very lowest level of biological complexity, the random generation of a genetic programme can be ruled out. The probability of this happening is as low as that with which shaking a box full of letters will lead to a textbook of biology.

As it is simply a fact that life nonetheless does exist on Earth, the molecular biologist Jacques Monod regarded himself as forced to interpret the origin of life as a "singular" event, one that was unique in the entire universe and will not be repeatable. As proof of this hypothesis he refered to the fact that the pattern of the genetic information carriers appear to be completely irregular. The lack of any law-like regularity, he argued, could only be explained by the random origination of those patterns at the threshold of life, even though the a priori probability of this happening is practically zero (see Sect. 2.5).

An equally one-sided, diametrically opposed, position was adopted in the 20th century by a number of prominent physicists. Niels Bohr, Eugene Wigner, Michael Polanyi, Walter Elsasser and others took the view that the existence of living matter was inaccessible to explanation within the known laws of physics and chemistry. Ultimately they developed—though from various different perspectives—the idea of the autonomy of laws governing living matter.

If we wish to subject the epistemological claim of such hypotheses to critical analysis, then the idea of natural law must be understood sufficiently broadly to allow it to be applied in a meaningful way to the problem of the origin of genetic information. Naturally, the same applies to the concept of chance as put forward by Jacques Monod. Both concepts can best be expounded within the framework of the so-called algorithmic complexity theory [17]. We shall first demonstrate this by considering the concept of chance.

In modern science the idea of chance is intimately associated with problems of computability. In mathematics, such problems are simulated with the help of the so-called Turing machine. For our problem, too, the idea of the Turing machine will prove helpful. Its principle is set out in Fig. 3.6. The basic parts of a Turing machine are an infinite tape with binary symbols and a tape head that can read the numbers on the tape and overwrite them with binary symbols. In each step, the tape head can read a single symbol, or print a single symbol out, and in doing so it can move the tape one position (corresponding to one symbol) to the left or to the right. The movement of the tape head is governed by a fixed set of transition rules that determine in which direction the tape head is to move, which symbol it is to print out and what "internal" state it is to adopt.

The Turing machine is to some extent an idealised computer, just as in physics the Carnot engine is an idealised heat engine (see Sect. 8.1). The Turing machine shown in Fig. 3.6 possesses only two possible states and can only carry out simple calculations. Turing machines with a tape head that can adopt more than two states

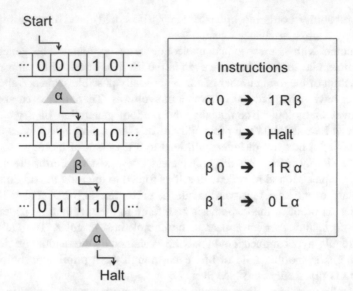

Fig. 3.6 Operation of a single-tape Turing machine. The machine operates on an infinitely long tape divided up into squares ("cells"), each containing a binary symbol (0 or 1). The tape is read by a monitor ("head"). In each step the monitor can read a symbol, or print one out, and it then moves one square to the *left* or the *right*. Moreover, the monitor can possess two internal states (α or β). The "programme" for movement of the monitor consists of a set of instructions; these determine, according to the four possible combinations of the monitor's state and position, in which direction (*left*, L, or *right*, R) the monitor is to move, which symbol (0 or 1) it is to print out and which internal state (α or β) it is to adopt. The illustration shows a short computation, in which the input "00010" is turned into the output "01110" and the programme stops after two steps of calculation

have a wider spectrum of transition rules. They are even universal, in the sense that they can simulate the behaviour of other machines that are much larger and more complex than they themselves are. In such cases, the complete logical state of the larger machine is represented on the input tape in coded form, and every step of the calculation performed by the larger machine is broken down into steps which are small enough to be simulated by the smaller machine.

By using the model of Turing machines, the idea of randomness can be expressed more precisely. Here a binary sequence is defined as being "random" if it can only be calculated (or generated) by a Turing machine which contains approximately as many binary symbols on the input tape as the sequence that is being generated. In short: for the calculation of random sequences the input programme has the same length as the output. In contrast, if the input programme is substantially shorter than the output, the generated sequence is defined as "ordered" (see also Sect. 6.3).

The fact that the degree of incompressibility of a sequence of symbols is a characteristic of its degree of randomness was pointed by a number of mathematicians [4, 15, 24]. They went on to equate the degree of randomness of a sequence with its degree of complexity. This makes sense, because the complexity of a sequence decreases in step with the discovery of shorter algorithms by which the sequence can be generated. As long as no shorter algorithm can be found the sequence must be regarded as inherently complex. But from the moment when a shorter algorithm is discovered the sequence must be considered—in retrospect— only as being complicated. Thus, the notion of complexity has an objective as well as a subjective aspect.

To express this relationship more exactly: The complexity K of a binary sequence S is given by the length L of the shortest algorithm or computer programme p with which S can be generated with the help of a computer C:

$$K(S) = \min_{C(p)=S} L(p)$$

The algorithmic definition of complexity has two interesting consequences. First, sequences of maximum complexity turn out to be random sequences. Secondly, the transition from ordered to random sequences is a gradual one. If a given sequence is ordered, this can also in principle be demonstrated. To do this, it is sufficient to find an algorithm for the generation of the sequence that is substantially shorter than the sequence itself. If, conversely, one wishes to demonstrate that a given sequence is random, then it has to be shown that there is *no* such compact algorithm, which means that the sequence is not significantly compressible. However, such a proof of the non-existence of an algorithm would run up against the same principal limitations as has been shown by Kurt Gödel in his response to Hilbert's idea of an all embracing formal system of mathematical proof [5] (see Sect. 2.2).

However, it is relatively easy to find out the *proportion* of random sequences in the set of all combinatorially possible binary sequences of length n. As the transition from ordered to random sequences is a gradual one, we first have to decide what degree of randomness we wish to detect. For example, if we regard all sequences with a complexity $K \geq n-10$ as being random, then the task is to detect all the sequences that cannot be compressed by more than 10 bits. For this purpose we have to count all ordered sequences that are of complexity $K < n-10$.

From this point of view there are 2^1 sequences of complexity $K = 1$, and 2^2 sequences of complexity $K = 2$, and 2^3 sequences of complexity $K = 3$, ..., and finally 2^{n-11} sequences of complexity $K = n-11$. Thus, the total number of ordered sequences of complexity $K < n-10$ is

$$\sum_{i=1}^{n-11} 2^i = 2^{n-10} - 2$$

As none of these algorithms (with $K < n-10$) can generate more than one binary sequence, there are fewer than 2^{n-10} *ordered* binary sequences. These make up a 2^{-10}th part of all n–membered binary sequences ($2^{10} = 1024$). This means that, among all n–membered binary sequences only about every 1000th is non-random and can (therefore) be compressed by more than 10 bits.

With the help of algorithmic complexity theory, far-reaching conclusions about the origin of genetic information can be drawn [17]. To retrace this, we must first translate the nucleotide sequence of a genome into the language of information theory.

Let us consider the genome of the RNA-virus MS2. As an RNA molecule is built up from four basic units, we need two binary symbols to encode any one of these four units. For example, with the binary coding

$$A = 00, U = 11, G = 01, C = 10$$

the segment of the genome of MS2 virus, shown on page 39, would look like this:

...1101100010011111101110100000100101110110111010110011010101...

Thus the question of the origin of genetic information turns out to be tantamount to the question of the origin of specific binary sequences of the kind shown above. Jacques Monod asserted that modern experimental and computational techniques for studying and comparing macromolecules taken from the living world do not provide any indication for the existence of a theoretical or an empirical rule that would allow one to predict, for example, the identity of the last unit in a n-membered chain molecule on the basis of the preceding n-1 known units [22]. Indeed, genetic research has so far failed to unearth any regularity, apart from short exceptions, that might suggest the presence of a hidden algorithm govering the physical structure of a genome. Nevertheless, we cannot infer from this that these sequences are truly random. We merely know that the great majority of all genetic sequences are random. However, we cannot know this of a particular given sequence. Here, there appear to be fundamental limitations placed upon what we can know.

What about the vitalistic hypotheses that assume the existence of autonomous laws applicable to living matter? To answer this question we must express the concept of "law" in such general terms that it also includes the notion of life-specific laws. Here again, algorithmic complexity theory comes to our aid. According to this, any sets of data that can be expressed as binary sequences possess law-like properties precisely when they display a regularity that makes them compressible. This applies not least to the kind of data that we acquire by observing Nature or by conducting experiments. In science, algorithms of this kind of regularities are ascribed to the action of "natural laws".

From an algorithmic standpoint the notion of natural law can be expressed so generally that it even embraces the—relatively undefined—idea of a "life-specific" law [17]. Concerning the origin of genetic information such a law could be understood as an algorithm guiding the specific arrangement of nucleotides in the genomes of living beings. Yet an algorithm of this kind can only be law-like in nature if the symbol sequences that it generates are not random, and if the algorithm itself is more compact than the sequences to which it gives rise.

It is true that the existence of compact algorithms, such as are postulated by vitalistic hypotheses, cannot be refuted, as their non-existence cannot be proven. However, on the other hand, vitalistic hypotheses have never provided any concrete indications that algorithms of this kind actually exist. Vitalistic hypotheses are merely apparent solutions which are based upon gaps in our physical and chemical knowledge of the world. As they are fundamentally not amenable to falsification, they belong to the category of pseudo-sciences.

So far, we have learned something about the possibilities and limitations of obtaining objective knowledge about the origin of life. From this, it is easy to imagine that we may, in the same way, learn something about the computability of the world in general. Here we must think immediately of an idea that has repeatedly enjoyed favour among physicists. It is the idea of an ultimate theory that embraces all natural laws in a general algorithm. Because of its universal applicability such a "world formula" would become a keystone of our scientific understanding of the world.

What forms, then, might the ultimate laws of Nature adopt? It has been argued, in the tradition of Kantian epistemology, that the ultimate laws must be fundamental in the sense that they formulate the conditions under which physical knowledge of the world becomes possible at all [26].

Within the framework of the algorithmic theory of complexity we can even present a more abstract and radical view of the "ultimate" laws of Nature. To that end we only have to make use of the fact that the vast quantity of observational and experimental data, which are obtained through scientific exploration of the material world, can be represented by an equally vast sequence of binary digits. Any regularity in this huge set of data that allows predictions of hitherto unknown data may indicate the action of some law. From this perspective, the search for a universal theory is equivalent to the search for the most compact algorithm that describes all regularities among the data, caused by the operation of natural laws. Such an irreducible algorithm is, by definition, a random sequence of binary digits.

At the first glance this result seems to be curious. Yet in fact it is a radical consequence of the idea that the "world formula" must involve everything that is law-like in the physical world, whereas the formula itself has no further law-like explanation. However, as mentioned above, the randomness of an algorithm and thus its property of being unamenable to further compression can never be proved. This immediately implies that the question of the completeness of physical theories has no definite answer. Any supposed completeness can only be refuted, namely by the discovery of a new and even more compact universal algorithm. Therefore, we will never know whether the end of physics has been reached or not.

In fact, science has no access to an absolute truth that would mark the final goal of scientific knowledge. From this point of view, the idea of a universal formula, by which all the phenomena of Nature can be computed, turns out to be only a guiding principle which, however, is indispensable for scientific progress. In exactly this sense, science is perpetually engaged in the search for ever simpler algorithms that allow us to compute the complex phenomena of the world.

3.3   How Does Information Originate?

The problem of the computability of the world concerns, not least, the riddle of the origin of life. For example, let us take up once more the statistical problem of the origin of genetic information. Our analysis so far has demonstrated that the probability of the spontaneous generation of information by pure chance is practically zero. However, this analysis set out from the tacit assumption that only a single sequence among the innumerable alternative sequences contains the information needed for the construction of a living system. Yet this assumption cannot be correct: in the course of evolution the genomes of organisms have been improved continually by random mutation and natural selection. There must therefore, inevitably, be vastly more sequences that carry meaningful—that is, functional—information than we have so far assumed. For a given sequence, we have no idea how many of the possible alternatives may carry such information. To judge this, we require detailed knowledge about the information space surrounding a given sequence. However, in view of the gigantic dimension of this space, it seems impossible to obtain such knowledge even for the smallest organisms. In that regard, it need only be remembered that even the genome of a simple RNA virus has already 10^{600} sequence alternatives (Sect. 3.2).

Since we do not have any knowledge about the information content of those sequences, we based our probabilistic calculation on the least favourable assumption. We assumed that among all the combinatorially possible sequences there is only *one* sequence encoding the information for a living being. Yet, if under this extremely pessimistic supposition we still succeeded in finding a solution to the problem of the origin of life, then this model would have the greatest explanatory power.

What might go into a possible answer to the question of how genetic information originated? Our considerations up to now have been restricted to asking how probable it is that a single chance event can lead to extremely complex information. However, other chance-based mechanisms are also imaginable. Instead of an all-at-once process one might imagine a gradual generation of information. In that case the process would start ab initio with the synthesis of a nucleotide sequence by adding stepwise random nucleotides to the growing chain until a predefined sequence is reached that carries some information for a living being. This process, however, presupposes a perpetual evaluation that operates in the following way: A nucleotide at a given position is only fixed in the growing chain when it is the same as a predefined nucleotide of the target sequence at that position. Otherwise the nucleotide occupying the position will be changed further, by trial and error, until the position is accurately occupied. On average (with four types of nucleotide), every fourth attempt will be successful, so that even a sequence of length comparable to that of a bacterial genome would be attained after a (realistic) number of some 16 million "attempts".

A process of stepwise optimisation is clearly highly superior to an "all-at-once" process. However, this process requires a selection among molecules that is very similar to that described by Charles Darwin. In fact, the theoretical und experimental investigations of this problem during recent decades have demonstrated that under certain physical conditions selection and evolution in the Darwinian sense are already active at the prebiotic level of matter.

For selection among molecules to take place, three prerequisites must be met. (1) The system must not be in thermodynamic equilibrium, since otherwise selective changes are not possible. This in turn means that from a physical point of view a selective system must be an open one: it must be able to exchange matter and energy with its surroundings. (2) The molecules must be capable of autocatalysis—that is, they must be able to reproduce themselves. (3) The self-reproduction must be "blurred", as only in that case is progressive evolution in the Darwinian possible. Thus, the *physical* definition of life as given at the beginning of this chapter is in accordance with the physical requirements of prebiotic selection and evolution.

The assertion that the statistical problem of the origin of life is solvable by selective behaviour of prebiotic matter can be verified by a simple computer simulation [17]. Here we can even dispense with the thermodynamic requirements that are necessary for natural systems. In this case self-reproduction and mutability are already sufficient to trigger the process of selection and evolution.

Making use of the analogy between the structures of genetic and linguistic information (see Sects. 3.1 and 5.4) we symbolize the functional content of a genome by a meaningful sentence of human language. In order to simulate most reasonably the initial conditions of the prebiotic scenario, we start with a random sequence of letters (such as might be obtained by repeatedly throwing dice):

ULOWTRSMILABTYZC

We now ask whether a pre-determined sentence, which represents the target of our experiment, can arise from a random sequence of letters by a Darwinian process of selection and evolution. It goes without saying that in Nature such a target sequence does *not* exist. But the problem we are investigating here is a purely statistical one: We are going to demonstrate that the problem of the origin of information could be solved in principle by a regular behaviour of matter within a reasonable time scale. The further question of how the "protosemantics" of genetic information could arise we shall discuss below.

To keep the experiment as clear as possible we choose as target-sequence some meaningful sequence which has the same length as the random starting sequence, for example:

EVOLUTION THEORY

We first convert this sequence (consisting of fifteen letters and one space) into a binary sequence. To encode all letters of the alphabet, including the space, we need

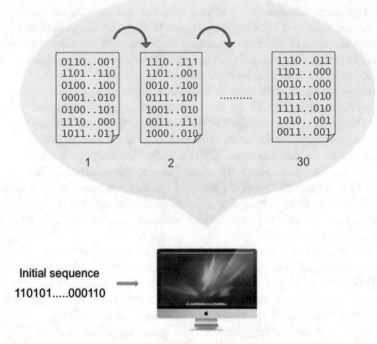

Fig. 3.7 Serial-transfer experiment simulated in a computer. A random starting sequence is continuously reproduced in the computer, whereby the reproduction is subject to mutation. By means of a special transfer method, selection pressure is applied to the reproducing sequences (for details see text). Under such conditions the starting sequence evolves until a predetermined target sequence is reached. The rate of evolution depends upon parameters such as the mutation rate and the reproductive advantage. The latter specifies the evolutionary progress that a sequence has made

at least five binary characters per letter. In binary coding, this sequence has already $2^{80} \approx 10^{24}$ alternatives. The probability that a random-number generator would produce a series of zeros and ones that corresponds exactly to our target sequence is virtually zero. Instead, we want to reach the target sequence by a process of selective optimisation. The experimental procedure we use, to exert selection pressure on our test sequence, is the method of "serial transfer" (Fig. 3.7).

The experiment starts by entering the random sequence into the computer. This is programmed to copy the binary sequence again and again, whereby each copy can itself be copied. In that way the property of *self-reproductivity* is build into the computer programme. However, the reproduction is not intended always to be error-free; rather, there is a certain built-in error rate in the copying, by which—now and again—a zero is replaced by a one, and vice versa. The occurrence of errors is taken to be completely random. In this way the biological phenomenon of *mutation* is simulated. As every sequence is self-reproducing, the mutant sequences will increase in number.

To allow a Darwinian process of evolution to set in, two further requirements must be met. There must be a selective evaluation of the mutant spectrum, and the system must be permanently under selection pressure. Both conditions can easily be realized. The evaluation of a mutant in the computer experiment is effected by means of its reproductive success (or failure) against the other mutants. More precisely: every sequence that is one binary symbol closer to the target sequence, shown above, is allowed to reproduce itself slightly faster than the parent sequence from which it was copied. Thus, as in real biological systems, the selection value of a mutant is defined by its "differential advantage" in reproduction.

What is happening under these conditions in the computer? Initially, the starting sequence is copied, and each copy produced includes random errors at a rate given by the pre-defined error frequency. To put the system under limitation of growth the population size must be permanently controlled. As soon as the population has reached a critical size of 100 copies the total number of copies is reduced—purely at random—to 10 copies. These are then allowed to grow back up to 100 copies, which in turn are again reduced at random to 10 copies. The procedure is now repeated over and over again, according to the same pattern, so that the size of the population of sequences, averaged over time, remains constant.

Under these conditions the mechanism of selective optimisation is relatively simple: At any moment, the distribution of self-reproducing sequences represents a "value level" that is given by the mean of the selection values of all sequences present. However, this distribution will change in time. According to the prevailing selection pressure, all sequences whose selection value lies below the mean selection value are excluded from further optimisation. In this way the mean value is inexorably displaced towards ever higher values, and this defines an "evolution gradient" along which the system automatically optimises itself.

Thus the selective optimisation of the system is based on a self adjusting value level, dominated by the sequence that is closest to the pre-defined target sequence and the mutant distribution at any moment. By selectively excluding those sequences that are less similar to the target sequence, the value level is raised, step by step, until it finally reaches that of the target sequence.

Figure 3.8 shows three phases of the experiment: the composition of sequences in the first, fifteenth and thirtieth generations of reproduction. In the thirtieth generation, selection equilibrium has been attained. It consists of correct copies of the target sequence and a stationary mutant distribution arising from this. As even the target sequence is occasionally reproduced incorrectly, it is always surrounded by a mutant distribution, which accompanies the target sequence like the tail of a comet.

The computer experiment demonstrates that the statistical problem of the origin of information can be solved in principle. Starting with a random sequence of letters, any target sequence can be reached in a relatively short time interval, by following the rules of Darwinian selection and evolution. This procedure, which is based on the interplay of chance *and* law, is much more efficient than an all-or-none process that rests entirely on the action of chance.

Nonetheless, the experiment has a fundamental flaw: it only works when a target sequence is defined in advance. According to the central working hypothesis of

1st transfer	15th transfer	30th transfer
ELWWSJILAKLAFTYJ	EUQLUDGON ?HEOQU	EVOLUTION THEORY
ELYWSJILAK?AFTYJ	ETOLUDGON ?HEOQY	EVOLUTION THEORY
ELWOSBCKEKLKUTIY	EUQLUDGONC?HEOQY	EVOLUTION THEORY
ELWOSBCKEKL!JTYY	EUOLUDGON LHEOQY	EVOLUTION THEORY
ELWOSBDKEKLAJTYY	EUOLUEDON LHEOQY	EVOLUTION THEORY
ELWWSJILAKLAFTYJ	EUOKUDGON LHE.QY	EVOLUTION THEORJ
ELWOSBCSEKLAJSYK	EUOLUDGON ?LUOQU	EVOLWTION THEORY
ELOWTBCKYKLIFTYJ	EUOLUDIONKLHEKQY	EVOLUDION THEORY
ELWWSJILAKL!FTYJ	EUOLUDGON ?HEOQY	EVOPUTION THEORY
ELOWTBCKZKLIJTYJ	EUOLUDGON ?HEOQY	EVO?UDION XHEORY

Fig. 3.8 Result of the computer simulation, described in Fig. 3.7. The distribution of sequences after various numbers of transfers is shown. By the 30th transfer, selection equilibrium has already been attained. It consists of the target sequence and the stationary mutant distribution arising from it

modern biology, however, no pre-defined targets exist in Nature. Does this objection mean that the simulation experiment fails to explain the origin of information from an informationless state?

The answer to this question depends on the perspective that one adopts. Considering a state in which information *is known*, the experiment clearly demonstrates that, starting from an informationless state, the target information can be easily reached by selection and evolution in the Darwinian sense. On the other hand, if a computer could generate meaningful information without any prior information, it would act like a perpetual-motion machine that generates information from nothing.

The explanatory difficulties that become visible here are well known in the philosophy of science. They are related to a fundamental theorem according to which any explanation is assumed to contain a potential prediction and vice versa. The symmetry of explanation and prediction is, not least, the basis for the principle of falsification that is essential in the methodological view of the exact sciences (Sect. 2.5).

However, as soon as random events begin to exert a decisive influence on the dynamics of natural processes, such as those of Darwinian evolution, the symmetry of explanation and prediction is disturbed. This was also the case in the simulation experiment above. Thus, the experiment represents only a simulation a posteriori. It cannot simulate the evolution of information a priori, because it necessarily requires the implementation of a value scale, determined by the target sequence, that directs the evolutionary progress.

This immediately prompts the question of where the value scale in natural systems comes from. Asking this question draws our attention to the "context" in which evolution takes place. According to the Darwinian theory of evolution the context is given by the environmental conditions of an organism, which are first and foremost biotic. In the prebiotic phase of evolution, when the transition from non-living to living matter took place, the context was given exclusively by the physical and chemical surroundings of the evolving matter. However, in contrast to the biotic environment, the abiotic environment did not contain any functional complexity that could have provided a target for evolutionary adaptation. Therefore, there must be mechanisms of evolution that are independent of environmental adaptation and yet are able to push prebiotic matter towards functional complexity.

A promising solution to this problem is offered by the idea of the "semantic code" [18, 19]. This concept has been developed within the framework of structural sciences (see Chap. 9) in order to obtain a rigorous approach to the *semantic* aspect of information. However, in contradistinction to the usual understanding of the notion of "code", the semantic code does not provide any rules for the assignment of symbols (or sequences of symbols) to another source of symbols. Instead, the semantic code is conceived as a value scale that a recipient applies to a piece of information that he wishes to "decode" with respect to its meaning [18].

This concept can also be applied to the generation of genetic information in prebiotic matter. Here, in absence of a biotic environment, the value scale that channels the evolution of matter towards the nucleation of life must be determined by general elements that are constitutive for any kind of functional order. These elements include co-operativity, effectiveness, hierarchical order, complexity and others. By contributing with different weights to the primordial formation of self-organizing systems the elements of the semantic lay down at the same time the individual character of the organization, which then may become the target of molecular selection and evolution in the Darwinian sense (for further details see [18]).

Returning to the simulation experiment we can draw the following conclusions: Although the explanatory capacity of the experiment is limited, it still has great heuristic value. It demonstrates that information as such can arise in a process of selection-based optimisation in the Darwinian sense. At the same time, the experiment visualises the vital role that is played by the context regarding the detailed content of information. However, this is not an obstacle that leaves a fundamental gap in our understanding of the origin of genetic information. On the contrary, it is a new source for a deeper understanding of the general principles that lie behind the early evolution of living matter.

A major aid to understand evolutionary optimisation also comes from the model of so-called "information space" (Fig. 3.9). This is basically a topological model, and it makes possible a detailed mathematical analysis of the complex interplay between mutation and evolution and the paths of evolution that emerge from them [10]. Roughly speaking, the optimisation process can be compared to a path leading through a multi-dimensional mountain landscape, the profile of which is determined by the fitness of the molecular species. The only constraint made upon the route taken by evolution is that it at all times must lead upwards. As the appearance of mutants with selective advantage is completely undetermined, evolution indeed follows a gradient; however, it never follows a pre-determined path to the next (local) maximum.

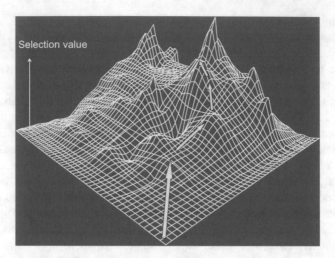

Fig. 3.9 Sequence space of a carrier of genetic information. In this very simplified illustration, each combinatorially possible sequence alternative of a given information carrier is represented by a point in the plane. If one plots above each point the selection value of the corresponding sequence, one obtains the so-called fitness landscape. The evolution of genetic information can then be described as an optimisation process that leads, on the fitness landscape, from a low (local) maximum to a higher (local) maximum. In reality, however, the topology of such landscapes is much more complicated than illustrated here. To get a true picture of the sequence space one has to take into account the neighbourhood relations among the sequences, the so-called Hamming distance (see [9]). Moreover, the selection value is a relative quantity that depends upon changes in the environmental conditions as well as changes due to evolutionary progress. In consequence, the fitness landscape will change with each evolutionary step

If the selection values depend only upon the physical conditions in the environment, then the structure of the information space remains the same as long as the environmental conditions remain unaltered. However, the assumption of a constant environment is an idealisation and it is never strictly met, even at the level of molecular selection systems: all individuals in a population of molecules contribute, by their physico-chemical properties, to the overall environmental conditions. For this reason, any displacement of concentrations within the population, induced by selection, leads of necessity to a change in the environmental conditions.

Moreover, the selection value of each species will also as a rule depend upon the numbers of all other species taking part in the selection process, so that each step in the process of selection and evolution will affect the topology of the information space. This means that the goal and the goal-directedness of evolution are inseparably and reciprocally related to one another, even at the elementary level of biological macromolecules. As the elementary processes that lead to evolutionary change are subject to chance, every process of evolution bears the stamp of its own historical uniqueness. This makes it clear that the Darwinian evolution model can only explain in a strict sense the origin of information as such, but not the final content of its evolution. The latter can only be explained a posteriori on the basis of plausibility considerations.

The information space in which the evolutionary optimization takes place is represented—even if only in very simplified form—in Fig. 3.9. This is because in reality, we are not dealing with a plain landscape as shown here, but with an n–dimensional space, where n is the number of all sequence alternatives of a given information carrier. Even for simple biological macromolecules these numbers exceed our imagination (see Sect. 3.2). Nevertheless, it is perfectly possible to investigate at least the general properties of such high-dimensional information spaces [10]. Such an analysis has, inter alia, revealed that optimisation in multi-dimensional spaces is substantially more efficient than in low-dimensional spaces. This finding has a plausible explanation: The larger the dimension of the information space is, the larger will be the number of possible pathways in the fitness landscape which in turn enhances the effectiveness of the evolutionary optimization.

Any extension of information space by random prolongation of the primary structure of biological macromolecules must therefore have played a decisive role in the origin of life. On the one hand, the extension of information space is tantamount to an increase in the *syntactic* complexity of potential information carriers, which in turn is a prerequisite for the nucleation and evolution of the proto-semantics of genetic information. On the other hand, an increase of the dimensionality of information space leads to an optimization of the optimization process itself, which thus becomes the decisive driving force for the evolution towards a higher functional complexity of living matter [19]. Even if these ideas are still "work in progress", they show how deep our insight into the nature of evolutionary processes has become, and how far our ability to compute the behaviour of complex systems has proceeded.

3.4   Evolution in the Test-Tube

The possibility that natural selection can take place among molecules seems already to have been suspected by Darwin. In a letter to the botanist Joseph Hooker he speculated: "But if (and oh! what a big if!) we could conceive in some warm little pond, with all sorts of ammonia and phosphoric salts, light, heat, electricity, &c., present, that a proteine compound was chemically formed ready to undergo still more complex changes, at the present day such matter w[ould] be instantly absorbed, which would not have been the case before living creatures were found."[1] At this point Darwin ends his speculation. However, the idea behind the fragment of thought that he reveals here is a broad vision of a chemical evolution, in the course of which matter organises itself. With it, Darwin laid down the basis for the

---

[1]Letter from Darwin to Joseph Hooker of 1871 (first published in [3]).

"primordial soup" scenario, which, in many different forms, has become a cornerstone of our present-day conception of the earliest phase in the origin of life.

Darwin's idea that a suitable reaction mixture, acted upon by various energy sources, should be able to produce and become rich in all the organic compounds that the living cell requires as basic chemical ingredients was taken up by modern science at the beginning of the 20th century [13, 23]. Finally, at the beginning of the 1950s, the chemistry student Stanley Miller performed a spectacular laboratory experiment, in which he succeeded in demonstrating the abiotic production of amino acids (the chemical building-blocks of the proteins) under conditions similar to those supposed to have prevailed millions of years ago on the Earth [20].

However, it must be noted that chemical evolution is not based on natural selection. On the contrary: under prebiotic reaction conditions, as simulated in Miller's experiment (and subsequent experiments by many others), everything is produced that the laws of physics and chemistry allow. For this reason, chemical evolution must have been a *divergent* phase of development, characterised by the emergence of a huge diversity of chemical compounds, which among others must have included the precursors of biological macromolecules. However, the decisive step towards the evolution of life must have been the *convergent* phase of material self-organisation, in the course of which biological macromolecules joined up into collectives of information-bearing molecules. The prototype of a system of this kind may have been a so-called "hypercycle" [8, 9] . This is a particular functional organisation of molecular information carriers that, owing to its co-operative behaviour, is able to overcome a fundamental information barrier at the threshold of life.

As we have seen, the action of natural selection among molecules is an indispensible part of any realistic picture of the origin of life, because it is only along this path that genetic information could originate. This raises the question of whether evolution in the Darwinian sense can in fact be observed in the test-tube. To answer this question the molecular biologist Sol Spiegelman and his staff [21] designed in the 1960s an experiment along similar lines to those of the computer experiment in Fig. 3.7 (in fact, the computer experiment was constructed later, along the lines of Spiegelman's experiment).

The test-tube experiment was conducted with the genetic material of the bacteriophage Q_β. This virus first penetrates its host cell and, once inside, begins to synthesize a virus-specific protein (termed "replicase") that causes the hereditary material of the virus, an RNA molecule, to be copied very rapidly. In the course of this the replicase moves along the viral RNA and catalyses its replication by reading off its nucleotide sequence base by base, and generating, by incorporation of energy-rich nucleotides, a new copy of the RNA.

Spiegelman managed to isolate the replicase from infected bacteria and to induce it to catalyse the replication of the viral genome in the test-tube, under controlled conditions. This technique for bringing a viral RNA to "life" outside its host cell opened up the opportunity to study in vitro the basic principles of the selection and evolution of biological macromolecules. The prototype of such experiments is illustrated in Fig. 3.10. Spiegelman's guiding thought in designing this

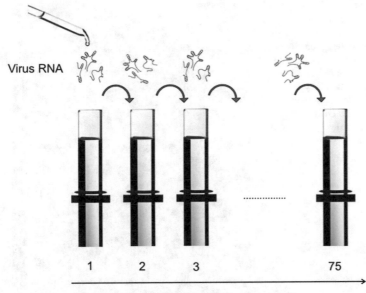

Decreasing periods of incubation

Fig. 3.10 Serial-transfer experiment with genetic material. By the dilution method shown, one can exert a selection pressure upon the Qβ replicase system (for details see text). The experiment is started by inoculating a reaction solution (test-tube 1) with the RNA of a Qβ virus. After incubation for a certain time, a sample of the reaction solution in test-tube 1 is transferred to fresh nutrient solution in test-tube 2. This procedure is repeated several times, and each period of incubation is made slightly shorter than the preceding one. In the original experiment the Qβ RNA was subjected to 74 such transfers. Spiegelman and his colleagues were able to isolate, as a final product, a variant of Qβ RNA that—under the "paradisiac" conditions of the test-tube world—had lost its genetic information and thus its infectiousness, while at the same time having become able to reproduce itself many times faster than the primary RNA [20]. Today, experiments of this kind can be performed efficiently by laboratory machines (see Fig. 3.11)

experimental technique was that under the ideal conditions in the test-tube most of the genetic material of the virus is superfluous, since all it needs to do is to be replicated. Spiegelman described his experimental approach by asking a picturesque question: What happens to the genome of the virus if nothing more is required of it than to follow the Biblical injunction "Be fruitful and multiply!"—as quickly as possible?

Spiegelman and his co-workers proceeded in the following way: First they provided the genetic material of the virus with everything it needed for multiplication in the test-tube (replicase, energy-rich nucleotides and so forth). They then incubated the reaction mixture, allowing multiplication of the RNA to take place. After a certain time interval they transferred a small portion of the mixture to another test-tube containing fresh growth medium for replication. Here, they again incubated the virus RNA to let it multiply. This procedure of dilution followed by incubation they repeated many times, always according to the same scheme, but

Fig. 3.11 Prototype of an evolution machine. The machine shown here was developed by Manfred Eigen and his co-workers at the Max Planck Institute for Biophysical Chemistry in Göttingen, Germany (photo: courtesy of the Deutsches Museum Bonn)

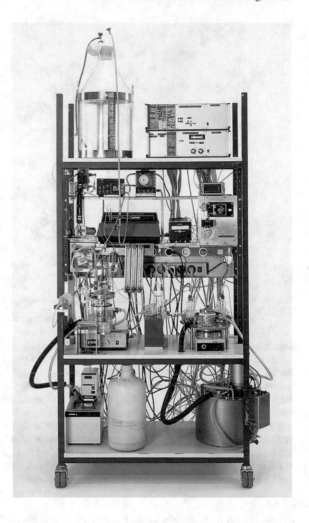

each time reducing slightly the incubation time between two successive dilution steps. In this way, they subjected the RNA molecules to an increasing selection pressure. For, as with each transfer the RNA molecules that replicated the fastest received an increasing differential advantage and an increasing chance, compared with the rest of the RNA population, of being present in the small portion transferred to the next tube of fresh growth medium.

Thus in Spiegelman's experiment every mutation that led to an increase in the effective replication rate of an RNA molecule must have been advantageous for that molecule. It is therefore no surprise that, toward the end of the experiment, the viral RNA had lost the greater part of its genetic information. Clearly, the genetic information that served the purpose of supporting the infection cycle is—under the luxurious conditions of the text-tube—no longer needed; in fact, it hinders rapid replication. Consequently, in that instance, evolution did not lead to an increase, but to a decrease in genetic complexity.

Following Spiegelman's experiments, the design and performance of experiments in evolution under controlled conditions were pursued in other laboratories, especially that of Manfred Eigen (see for example [16]). Today, these experiments can even be carried out automatically by computer-controlled machines (Fig. 3.11).

Apart from yielding new insights into the mechanisms of evolution, these experiments also laid the foundations for a novel kind of "evolutionary biotechnology", one that goes far beyond traditional biotechnology. The spectrum of its applications extends from the design of new medicines and vaccines through to the development of catalysts for industry.

The advantages of this technology are obvious: (1) In artificial evolution, the processes of multiplication and selection—unlike in natural evolution—can be uncoupled from one another. (2) Selection can be directed, as the mutation rate, the speed and the direction of artificial evolution can all be adjusted. (3) The selection of a biomolecule no longer requires the functional environment of a cell or an organism; it can take place under arbitrarily chosen, cell-free conditions. (4) In principle, one can set out from any point in genetic information space and, from there, begin a process of artificial evolution.

By means of artificial evolution, biomolecules can be produced that would have no chance of surviving in free Nature, and therefore never could be obtained in that way. Similarly, it will prove possible to optimise biomolecules that already exist. Within the framework of this technology, the mechanisms of natural evolution are employed; however, the products themselves are not subject to natural selection, but rather to the criteria of artificial selection and optimisation, and to these alone. Evolutionary biotechnology has no need of "natural" material or of the functional apparatus of living cells or organisms. It is simply applied molecular evolution under the controlled and reproducible conditions of the laboratory.

3.5 Are We Alone in the Universe?

The experiment by Stanley Miller was the bugle call for modern science's hunt for the origin of life. It set off a spate of investigations into the evolution of organic molecules. Not the least of these were studies, by new methods in radio astronomy, of chemistry in interstellar space. This was of interest because it was assumed that reaction conditions there might allow the abiotic formation of large organic molecules. Indeed, astrophysicists very soon came up with results, and today it is taken for a fact that a multitude of complex organic compounds are indeed present in interstellar space. Is it then not possible that, in the depths of space, there might exist Earth-like planets on which the conditions for life are also met? And, if life really emerges with law-like inevitability under the right conditions, is it then not equally inevitable that the sky over our heads is teeming with life? Some radio astronomers are convinced that in our Milky Way alone there are more than

100,000 planets on which life could develop to the level of intelligent civilisation [7]. Seen from that perspective, even the Milky Way seems hopelessly overcrowded.

At the time when the first organic molecules in space were detected, most exobiologists were convinced that life must have originated more than once in the universe, and that, with a probability bordering upon certainty, intelligent civilisations were also present. Thereupon a feverish search for intelligent extraterrestrial life commenced—a search that however had little to do with serious science and seemed more reminiscent of the gold rush of the 19th century.

The results of exobiological research also put new life into the old hypothesis according to which life reached our planet, in whatever way, from the depths of space. We find this idea, in primitive form, already in the ideas of antiquity. But at the beginning of the 20th century is was again taken up, by the physical chemist Svante Arrhenius, and dressed in pseudoscientific clothes. Since then, the idea of panspermia has fired the imagination of scientists and has found prominent advocates up to our time (see for example [14]).

If life on Earth is indeed the result of a cosmic infection, if life so to speak has fallen from the heavens, then of course the answer to the question of whether extraterrestrial life exists is a clear affirmative. It is equally clear, however, that the idea of panspermia (whether directed or not) leaves completely unanswered the question of the conditions for, and the possibility of, the existence of terrestrial and of extraterrestrial life; the question is simply banished to the depths of the universe.

If we wish to understand seriously the problem of the origin of life, then we are forced, like it or not, to consider it under the conditions of our *own* planet. All hypotheses or theories that seem to increase the chances of the existence of life by appeal to the unbounded size of the universe ultimately trivialise the problem. Every probability, however small, can be boosted to an arbitrary extent, and thus brought into the region of certainty, if the scene set for the event in question is chosen to be large enough.

However, we have no need of pseudo-explanations of this kind. As we have seen, the origin of life can be understood, at least in principle, as a law-like event. Yet one cannot conclusively infer from this that the universe contains diverse forms of life as we know it. A living system is not only the expression of universal natural principles, ones that apply to every possible form of life; living matter (and that is the decisive point) also has a long history of development, and it is this that confers upon each life form its specific appearance.

All organisms, from bacteria to humans, are unique products of evolution, insofar as their genetic programmes each represent a unique selection out of a virtually infinite plenitude of alternatives. This means that evolution possesses a practically unlimited range of possibilities, and it is completely out of the question that any two of these, starting from the same point, would follow the same course. If life has been sparked off several times at various places in the universe, then it is as good as certain that subsequent extraterrestrial evolution will not have followed the same course of development as on Earth. Even if the nucleation phase may be uniform to a certain extent the evolutionary pathways will finally diverge to the same degree as the complexity of life and its dependence on historical contingencies increases.

However, we have no idea of where the path of extraterrestrial evolution might lead and what the products of such an evolution might be. Although the numerical extent of the information space, and hence of the possible states that evolution can pass through, are known, its qualitative properties remain terra incognita. It will never be possible, either though observation or by theory, to obtain a complete picture of the material manifestations that are contained within the innumerable variants of a genetic programme. Moreover, as we are ultimately concerned with the question of the possible content of information, statistical methods are also of no help. Even if one could find thousands of sequences that carried meaningful information, then averaging these would tell us absolutely nothing about the information content of the remaining sequences not yet analysed. In short: Semantic information can just as little be subjected to statistical analysis as a telephone book can be characterised by the average of the telephone numbers that it contains.

One sometimes hears the view expressed that evolution, precisely because its course is irreversible, channels its own development and in that way restricts the number of possible paths along which it might proceed. Behind this view lies the idea that the existence of a large number of developmental paths is no longer possible once evolution has set out along a particular one. This thought is right, beyond doubt, and it may be the reason why all organisms, different as their development may have been, still resemble one another in many respects. It may also be the reason for an observation made by the palæontologist Simon Conway Morris which seems to indicate that evolution has invented the same mechanisms several times, each time independently of the others [6].

Whether, and to what extent, evolutionary developments along such convergent paths have taken place, is an as yet unanswered question. However, it seems reasonably certain that evolution, even when it sets out twice from the same starting conditions, never takes the same or even a similar course. The process of evolution is as little reproducible as it is reversible. In other words, it is a typical historical process. This is unaffected by the fact that life on another planet might well makes use of the same chemistry, based upon nucleic acids and proteins, as on Earth. That the same chemistry will lead automatically to similar life forms is just as little plausible as the assertion that two chemical factories which use the same raw materials necessarily manufacture the same products. On the contrary, it is to be expected that two evolutionary processes starting at different places in the universe will lead to genetic programmes with fundamentally different content.

This raises again the question of the fundamental quality characterising living beings. We have seen repeatedly that it is impossible to develop a complete concept of what life is without ending up in a tautology. Unfortunately, we only know a single form of life: the one that developed on our planet. We therefore have great difficulty in developing a picture of life independent of our own, and such pictures are so abstract and diffuse that they lose touch with reality. Therefore it is not surprising that in our imagination the extraterrestrial beings are also "human": everyone knows the ugly but lovable creatures that have already invaded the

Earth—as Extraterrestrials (E.T.), Alien Life Form (ALF) or green men from Mars —by having taken over our media.

References

1. Bohr, N.: Light and life. Nature **131**, 421–423 (1933)
2. Butler, P.J.G., Klug, A.: The assembly of a virus. Sci. Am. **239**(5), 62–69 (1978)
3. Calvin, M.: Chemical Evolution. Clarendon Press, Oxford (1969)
4. Chaitin, G.: On the length of programs for computing finite binary binary sequences. J. Assoc. Comput. Mach. **13**, 547–569 (1966)
5. Chaitin, G.: Information, Randomness & Incompleteness. World Scientific, Singapur (1987)
6. Conway Morris, S.: Life's Solution. Cambridge (2003)
7. Drake, F., Sobel, D.: Is Anyone Out There? Delacorte Press, New York (1992)
8. Eigen, M.: Self-oganization of matter and the evolution of biological macromolecules. The Science of Nature (Naturwissenschaften) **58**, 465–523 (1971)
9. Eigen, M., Schuster, P.: The Hypercycle. Springer, Heidelberg (1979)
10. Eigen, M.: From Strange Simplicity to Complex Familiarity. Oxford University Press, Oxford (2013)
11. Fiers, W. et al.: Complete nucleotide sequence of bacteriophage MS2 RNA: primary and secondary structure of the replicase gene. Nature **260**, 500–507 (1976)
12. Fraenkel-Conrat, H., Williams, R.C.: Reconstitution of active Tobacco Mosaic Virus from its inactive protein and nucleic acid components. Proc. Natl. Acad. Sci. USA **41**(10), 690–698 (1955)
13. Haldane, J.B.S.: The origin of life. Ration. Annu. **148**, 3–10 (1929)
14. Hoyle, F., Wickramasinghe, C.: Evolution from Space. Simon & Schuster, New York (1981)
15. Kolmogorov, A.N.: Three approaches to the quantitative definition of information. Probl. Inf. Trans. **1**, 1–7 (1965)
16. Küppers, B.-O.: Towards an experimental analysis of molecular selforganization and precellular Darwinian evolution. The Science of Nature (Naturwissenschaften) **66**, 228–243 (1979)
17. Küppers, B.-O.: Information and the Origin of Life. MIT Press, Cambridge/Mass. (1990). [Original: Der Ursprung biologischer Information, 1986]
18. Küppers, B.-O.: Elements of a semantic code. In: Küppers, B.-O. et al. (eds.): Evolution of Semantic Systems, pp. 67–85. Springer, Heidelberg (2013)
19. Küppers, B.-O.: The nucleation of semantic information in prebiotic matter. In: Domingo, E., Schuster, P. (eds.): Quasispecies: From Theory to Experimental Systems, pp. 23–42. Springer International Publishing, Switzerland (2016)
20. Miller, S.L.: Production of amino acids under possible primitive earth conditions. Science **117** (3046), 528–529 (1953)
21. Mills, D.R., Peterson, R.L,.Spiegelman, S. (1967). An extracellular Darwinian experiment with a self-duplicating nucleic acid molecule. Proc. Natl. Acad. Sci. USA **58**, 217–224 (1967)
22. Monod, J.: Chance and Necessity. Vintage, New York (1971). [Original: Le Hasard et la Nécessité, 1970]
23. Oparin, A.I.: The Origin of Life on the Earth. Academic Press, New York (1957). [Original: Proiskhozhedenie Zhizni, 1924]
24. Solomonoff, R.J.: A formal theory of inductive inference. Information and Control **7**, Part I, 1–22; Part II, 224–254 (1964)

25. Weizsäcker, C. F. von: Foreword to Küppers, B.-O.: Information and the Origin of Life, pp. xi–xv. MIT Press, Cambridge/Mass. (1990). [Original: Küppers, B.-O.: Der Ursprung biologischer Information, 1986]
26. Weizsäcker, C.F. von: The Structure of Physics. Springer, Heidelberg (2006). [Original: Aufbau der Physik, 1985]

Chapter 4
What Is Information?

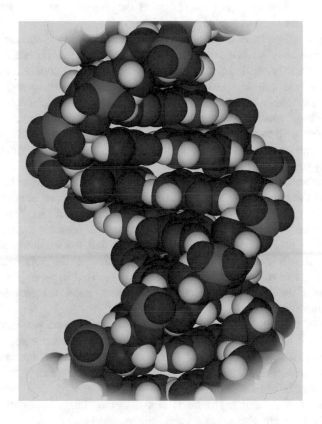

According to modern biology, the molecules of heredity, such as the DNA model depicted here, carry the information for a living being. Yet can molecules really carry information at all? Is information not inevitably associated with human communication? Is information, as biology understands it, actually a natural entity? What is the true nature of information?

© Springer International Publishing AG 2018   67
B.-O. Küppers, *The Computability of the World*, The Frontiers Collection,
https://doi.org/10.1007/978-3-319-67369-1_4

4.1   Dimensions of Information

Information is based upon symbols and sequences of symbols. These may be rich in content, that is, they may express some kind of "sense" or "meaning" for a potential recipient. This content can in turn take effect in various ways by prompting an action or a reaction on the part of the recipient or in the recipient's surroundings.

Consequently, we can distinguish three dimensions of information. The *syntactic* dimension denotes the way in which the symbols are arranged, as well as their relationship one to another. The *semantic* dimension includes the relationships among the symbols and what they mean. Finally, the *pragmatic* dimension includes the relationships between the symbols, what they mean, and the effect that they engender with the recipient.

However, the dissection of the concept of information into syntax, semantics and pragmatics can only be justified in the context of its scientific analysis and systematisation. Strictly speaking, the three dimensions form an indissoluble unit—not least because one cannot speak of syntactic, semantic or pragmatic aspects of information separately without referring at the same time to the other two dimensions and taking account of them. Moreover, besides the pragmatic dimension, further aspects of information (such as its novelty, complexity and other aspects) may also contribute, albeit in particular cases with different weights, to the semantics of information (see Sect. 5.3).

Take syntax first of all. This only attains any kind of meaning when the semantics of the symbols as such and their relationship to one another are predefined. Semantics, in turn, presupposes a syntactic structure as "carrier" of the semantics, even if—of course—the semantics go beyond their mere syntax. The same is true of pragmatics. The reason for this is that the meaning of an item of information fundamentally become accessible through the relationship of the symbols to the external world. This is ultimately realised in an objective sense by the effect that the item of information engenders upon the recipient and the recipient's surroundings.

Because of their intimate interrelationship, semantics and pragmatics are frequently lumped together into the so-called semantic-pragmatic aspect of information. Alongside the pragmatic aspect, however, there are further facets of semantic information, ones that are expressed in the novelty, complexity, selectivity and other aspects of information.

In a manner of speaking, semantics represent the "hinge" between syntax and pragmatics; one might also say: between the structural and the functional aspect of information. Even if a function always presupposes structure(s) as carrier(s) of the function, functions cannot be reduced to their structural carriers, as there is clearly no law-like connection between these two levels. An example of this might be human language, in which the meaning—and thus the pragmatic (or functional) relevance—of words and sentences is fixed by the syntax, but is still not reducible to the syntax.

The arbitrary assignment of the semantics to the syntax raises the acutely difficult question of whether the semantic dimension of information can ever become the subject of an exact science based upon abstraction, idealisation and generalisation. After all, in human communication the exchange of meaningful information is always an expression of the mental activity of human beings, with all their individual and unique attributes. Can these be blanked out, so that the essential, general characteristics of semantics—as far as any such exist—can be crystallised out? Is there possibly a semantic code, according to which meaningful information is constituted? In the face of questions like this, which seem to overstep the boundaries of exact science, it is easy to understand why the founding fathers of information theory initially sought a measure of information that was independent of the content of the information.

The first steps towards such a theory, which abstracts from the semantics of information, emerged from the work of communications engineers. In fact, communications technology is not concerned with the actual content of an item of information; rather, its primary task is the mere transmission of a given sequence of characters from the sender to the recipient. This demands that the transmission route, the so-called information channel, works as accurately as possible, so as not to corrupt the sequence of characters on its way to the recipient.

Taken on its own the reliable transmission of a message is an important technical challenge; but this problem has nothing whatever to do with the content of the message being conveyed. It is true that communications engineers also speak of the "information content" of a message. However, in doing so they are referring not to the actual content, but rather to the sheer "quantity" of information, defined as the number of characters that are to be transmitted. Warren Weaver has expressed the technical understanding of information in the following way: "In fact, two messages, one of which is heavily loaded with meaning and the other of which is pure nonsense, can be exactly equivalent, from the present viewpoint, as regards information" [14, p. 8].

Obviously, information has the task of dispelling uncertainty. It is for that reason that the information content of a message seems the greater to us, the less we expected that particular message—in other words, the greater its novelty value for us is. This thought also lies behind the classical concept of information, which is known as Shannon's information theory. According to that theory, the probability p_i with which a given message (symbol sequence) x_i out of a given set of possible messages will arrive, is a simple measure of the information content of the message. More exactly: the information content of a message x_k is inversely proportional to its expectation value p_k, that is:

$$I_k \sim 1/p_k$$

Claude Shannon made a further requirement of the measure of information: being an extensive quantity, i.e. one that increases linearly with the amount, it had to be additive. However, probabilities per se do not have this property. For this reason Shannon was forced to replace the probabilities by their logarithm.

$$I_k = \text{ld}\,(1/p_k) = -\text{ld}\,p_k \quad (\text{ld} = \text{logarithm to the base 2}).$$

With this little trick the measure of information becomes additive. Moreover, with the introduction of binary logarithms, the definition also moves a step in the direction of the binary coding that is usual in communications engineering and computer technology.

The information measure proposed by Shannon, which relates the information of a message to its novelty value, makes up the basis of the information theory used in communications engineering. In the simplest case: if a message source generates N messages, all of which have the same expectation value for the recipient, then the recipient would need to make

$$I = \text{ld}\,N$$

binary decisions in order to select a particular message. The quantity I is therefore also termed the "decision content" of the message in question. In fact, this simple measure had already been proposed by Ralph Hartley [5] many years before the development of modern information theory started.

The probabilities p_k are already defined by the recipient's prior knowledge. However, they are also objectively knowable quantities for all recipients who fulfil the same presuppositions and have at their disposal the same methods of gaining knowledge. One might say that the information measure developed by Shannon is subject-related in an objective way [15]. This expression already implies that information in an absolute sense does not exist. Instead, information is always to be seen in relation to the recipient. This aspect, which is inherent in any kind of information, is also denoted as the "context dependence" of information.

Shannon's measure of information refers to a source of messages which is described by a probability distribution $P = \{p_1, \ldots, p_N\}$. A characteristic of this message source is its average information content given by the statistical weight

$$H = \sum p_k I_k = -\sum p_k \text{ld}\,p_k$$

The quantity H is also referred to as the entropy of the message source, because it has the same mathematical structure as the entropy function in statistical mechanics (see Sect. 7.2).

By additional observation the probability distribution

$$P = \{p_1, \ldots, p_N\} \text{ with } \sum p_k = 1 \text{ and } p_k > 0$$

may change to

$$Q = \{q_1, \ldots, q_N\} \text{ with } \sum q_k = 1 \text{ and } q_k \geq 0$$

The associated gain in information can be expressed as the entropy difference:

$$H(Q)-H(P) = \sum q_k I_k - \sum p_k I_k$$

It is minor blemish that this definition may have positive as well as negative values, which in turn would contradict our everyday notion of a "gain" or an "increase" of information. But this apparent contradiction can be resolved by considering not the difference between two entropy values, and thus the difference between two average values, but rather by calculating the gain of information related to each single message and then calculating the mean value:

$$I(Q \mid P) = \sum q_k [I_k(p_k) - I_k(q_k)] = \sum q_k \mathrm{ld}\,(q_k/p_k)$$

The information gain, as calculated in this way by Alfréd Rényi [12], always fulfils the condition

$$I(Q \mid P) \geq 0$$

Even though the information theory that arises from communications engineering excludes any reference to the content of information, it still points the way, if only indirectly, to a particular aspect of its semantics, namely that of the novelty value of information.

The aspects of information described so far always refer to a source of messages that is characterised by a probability distribution. This raises the question of whether the information content of a message can be defined without making reference its source. The basis for such a measure of information would then be not the probability distribution, but rather the structure of the message itself.

In fact, a measure of this kind has been developed within the framework of the algorithmic theory of complexity. Here, the algorithmic notion of information rests entirely on the complexity of a message, which is defined by the degree of its incompressibility. We have made extensive use of this idea in Sect. 3.2. However, like the classical information theory of Hartley and Shannon, the algorithmic theory makes no reference at all to the content of information.

4.2   Naturalism Versus Constructivism

Shannon information and algorithmic information are the two pillars of the information concept in the natural sciences. For the theoretical foundations of biology they are in fact indispensable (see Chap. 3). On the other hand, information is first and foremost a central element of human communication. Therefore, it is justified to ask critically whether it is reasonable to make use of the concept of information with regard to the material phenomena of Nature. Not a few scientists regard the concept of information as a purely cultural item, one that at best may be used as a helpful metaphor in the description of Nature, but they deny absolutely that information is an objective property of matter.

The use of the concept of information in science has been criticised with particular vehemence by those philosophers of science whose thinking is very close to a philosophical school known as constructivism. This philosophy represents one of the most important epistemological currents in present-day philosophy. The numerous different forms in which it strides the philosophical stage are already in themselves impressive. The scale of constructivistic schemas covers the spectrum from the methodological constructivism of the so-called "Erlangen School" through the constructive realism of the "Vienna School" to the radical constructivism propounded especially in the "American School" of cognitive psychology.

All these philosophical approaches are based upon the central thesis that reality is constituted by the process of recognition itself even if they differ in their claims to generality. The experienceable and recognisable reality is, in that view, always a reality that the subject constructs and objectifies according to his own criteria. In its most radical formulation, constructivism goes so far as to assert that reality is *invented* and not *discovered* by the subject. In that way, constructivism inevitably takes up a position opposite to the so-called realism. The latter concedes that reality has its own mode of existence. This is to be understood in the sense that the existence of the objects of cognition is independent of the manner in which they are recognised. In other words: realism postulates that all knowledge ultimately proceeds from, and is conditioned by, the object of cognition. Thus, the clash between constructivism and realism primarily concerns the basic question of the weight given to subjectivity and objectivity in the process of gaining knowledge.

As naturalism is the predominant doctrine of the natural sciences these must be challenged above all by the constructivistic criticism of the scientific knowledge. Conversely, philosophical concepts that not only question the objectivity and validity of scientific knowledge, but also claim for their part to provide a strict foundation for the exact sciences, must be subjected to the most rigorous scrutiny in the corridors of science itself. In its dogmatic and polemic variants constructivism resembles to a certain extent the philosophical movement of the 18th century which also accused the research of natural sciences of being "blind" and "unimaginative" (see Sect. 1.4).

In the following paragraphs, I shall examine as an example the harsh criticism that has been put forward by the philosopher Peter Janich with repect to the widespread use of the concept of information in the natural sciences, in particular in biology [7, 9]. The analysis will make clear that the dogmatic application of constructivism to modern science inevitably leads to serious misunderstandings, if one neglects the historical roots of scientific ideas and concepts.

The target of Janich's criticism is the influence of the contempory analytical philosophy on modern science. This kind of philosophy—that is the central point of critisism—does not critically reflect the methodological basis of scientific research. Instead it restricts itself to the mere observation, analysis and description of the norms that guide the processes of scientific discovery without worrying about the origin and justification of such norms [8]. With respect to the "uncritical" methodological self-conception of modern science the constructivitic side refers to the philosophical debates of earlier times, when thinkers like Kant and others strove to find a foundation for the presuppositions and conditions of human knowledge. Precisely that is what *methodological* constructivism claims for itself: the ability to extent the epistemology and metatheory of science normatively, in the above sense.

However, methodological constructivism—and this is its particular signature feature—has a specific understanding of the "norms" that control the acquisition of knowledge. In fact, these norms are understood as *prescriptive* statements that specify the facts and circumstances in which the objects of science are constituted in a methodologically controlled form. This already lays bare the character of methodological constructivism: its rootedness in models of cognition, which arc conditioned and structured by human actions. Or as Janich puts it: "The rationality of knowledge can be reconstructed as the rationality of acting in order to obtain knowledge" [8, p. 5]. Accordingly, the theory of human action makes up a central part of methodological constructivism with the aim of delineating the normative, the objects of cognition constituting parts of science.

From this viewpoint the objects of our knowledge in science and in the everyday world are "constructions", inasmuch as they only become objects of our knowledge through our dealing with them in actions and discourse. As, however, the natural sciences tend rather to adopt the opposite standpoint of realism, Janich feels obliged to accuse the natural sciences of making inadequately clear the distinction between scientific research as a cultural action on the one hand and Nature itself on the other. According to Janich this inevitably leads to naturalistic fallacies. Therefore, he claims to confront the naturalism of the natural sciences with a culturalistic consideration of reality, one that unmasks the objects of knowledge of natural science and reveals them for what they in actual fact are: cultural constructs that arc contingent on human actions.

In order to differentiate this position more sharply from that of "naturalism", Janich characterises his own interpretation of constructivism as "culturalism". On the basis of this philosophical premise he proceeds to root through the various scientific disciplines and subject them to an epistemological critique. Thereby his arguments always follow the same pattern. First, science, above all natural science, is accused of working with inadequate definitions, confused concepts and explanatory deficiencies. Then the cause of this evil is traced back to the allegedly limited view of analytical science, which only has an eye for what is actually there and has a completely distorted perspective on the cultural roots of science.

In the following paragraphs I wish to demonstrate that the constructivistic critique of the analytical science is untenable. In doing so I will focus on the controversy regarding the concept of information in science, in particular in biology. But let me say from the start that I do not reject constructivism lock, stock and barrel. Quite on the contrary: the view according to which—put briefly—our scientific knowledge is also, in many respects, conditioned by our engendering actions within the context of our lifeworld, seems to me to be obvious. What I wish to criticise, however, is the exaggerated claim to providing a philosophical foundation of science that sets out from a narrow and dogmatic interpretation of the cultural roots of science according to which scientific knowledge must be exclusively ascribed to human actions and their presuppositions in our lifeworld.

Where can we place a lever to jolt the excessive claims of methodological constructivism back into place? The answer can be looked for in the writings of the constructivists themselves. For example, Janich admits that even a normative theory of scientific knowledge cannot simply be built up on the foundation of freely invented, or constructed, postulates. Such a theory, he argues, must take account of the practice of gaining knowledge, in science as in everyday life. For this reason "on the one hand, an adequacy condition must be fulfilled, according to which the common understanding of knowledge and discovery must be taken on board and 'reconstructed'", and on the other hand we "must preserve the achievements that become possible by recognised bodies of scientific knowledge—for example, the technical, prognostic and explicative achievements of the natural sciences" [8, p. 2].

Put straightforwardly, this means that the normative elements demanded by constructive theory of science are not evident in themselves, but are subject to certain checks and balances, ones that Janich refers to by vague formulations such as "common understanding" of knowledge and "recognised bodies" of scientific knowledge. Pronouncements of this kind make it clear that constructivism is not really in a position to fulfil its claim to providing a normative foundation of scientific knowledge. On the contrary: ultimately one is forced to appeal to the natural sciences' faculty of consensus. This is in fact characteristic of the procedure of normative foundations. The philosophical problems that arise in this connection have already been critically discussed elsewhere [1].

So what property of the "recognised" bodies of knowledge could one unanimously appeal to? There is no question that the human being is not only a cultural, but also a natural being. However, this emphasises the fact that some determinants of human action bear the imprint less of mankind's cultural history and more of its natural history. Yet how can we justify any knowledge of the natural determinants of human action in a constructivistic sense, if the theoretical basis for this justification itself depends upon the determinants that are to be explained?

Constructivism avoids this looming circular argument by not even considering the issue of the evolutionary and biological determinants of human action. Its adherents believe that they can duck this problem by making the laborious, and by no means convincing, attempt to demarcate human action from biological behaviour. Janich, for example, makes the following distinction: "One can call upon a person to act, but not (in the narrow sense) to behave. Actions one can refrain from —but not behaviour patterns. Actions can succeed or fail—but not behaviour patterns" [8, p. 7].

However, the criteria for distinction brought to bear by Janich become problematical when we consider not only closed, but also open behaviour patterns. Such open patterns, as observed mostly in living beings with higher degrees of organisation, can by their appearance hardly be distinguished from actions if one applies Janich's criteria. They show all the features of flexibility, indeterminacy, selectivity and purposiveness that are (also) characteristics of human actions [11]. The constructivist lobby will stalwartly emphasise the culturally determined law-like properties of action that determine the nature of knowledge. But this statement, one may reply, can be extended to include the evolutionary thesis that our forms of knowledge have been influenced by the patterns of behavour and experience that enabled early humans to survive in a hostile environment (see [4]).

The appeal to natural history clearly strikes at the Achilles' heel of constructivism; this is easily seen in the stilted attempts that Janich [6] makes to undermine the validity of hypotheses from natural history, or at least to make them dependent upon constructivistic validation procedures. He claims in this context that the uniqueness of natural events stands in contradiction to the repeatability of observations in the natural sciences, and that only knowledge of the causal connections in the form of technological know-how makes hypotheses about natural history possible. However, this argument is only correct to a limited extent, since the discoveries of modern physics have shown that even a law-like behaviour of matter does not necessarily lead to the repeatability and thereby the predictability of events (Sect. 2.2).

Admittedly, this subtle difference only becomes clear when we express the concept of causality very precisely. To do this, we first have to distinguish between a causal principle and a causal law. The causal principle, in its general form, merely asserts that there are no un-caused events, that every event is associated with a cause on which it is conditional. The nature of this connection is then stated in the form of general causal judgements that we call natural laws. In this way natural laws, for example, are founded upon a particular kind of causality, according to which similar causes always evoke similar effects. Only this kind of causality engenders the regular behaviour of natural phenomena and leads, in laboratory experiments, to approximately reproducible natural processes which in turn may become the basis of a technological know-how (see also Sect. 8.2).

However, research into complex systems has shown that in natural events there are certain non-linear connections between cause and effect. Consequently, similar causes can have quite different effects. This kind of causality leads to unpredictable, irregular behaviour as is typical of historical processes. The development of such systems may in principle be experimentally reproducible, but only if the experimental starting conditions can be reproduced with arbitrary exactitude—a requirement that is unfulfillable both in the laboratory and in free Nature. In the light of this, it is no longer possible to hold onto the simple idea that the laws of Nature are always propositions of experimental science "for which an artificially produced arrangement can lead to their realisation, and hence their testing" [6, p. 116].

The sense of the conclusion reached by Janich is the assertion that the laws of Nature are always a theoretically formulated capability to act and, therefore, represent factually a technical know-how. However, a strong indication of the limitations of this view is given by the fact, that—in step with the movement of the natural sciences towards the analysis of complex systems—the traditional laboratory experiment is increasingly being displaced by computer simulations. Indeed, computer simulations have become an independent, scientific, methodological tool with the help of which we can investigate the complex natural processes that are inaccessible to classical laboratory science.

There are numerous examples of this; let us consider just two such. In meteorology, weather processes can certainly be described with the help of known physical and chemical laws. But they can be investigated in laboratory experiments only to a limited extent. Instead, they are simulated by computer experiments, even though in many of these processes the actual course of events is unique. It was not least from computer-assisted meteorology that the rediscovery of, and research into, deterministic chaos received its impetus. This experience shows us more than clearly that knowledge of the natural laws does not necessarily mean that these can at once be turned into technology. Even if we understand quite well the law-like principles governing the atmosphere, this still does not automatically give us the ability to generate reproducibly meteorological states and processes.

A second example of the fact that scientific knowledge does not necessarily lead to technological capabilities is provided by the fundamental laws of thermodynamics. These laws belong without any doubt to the basic principles of Nature. The first law ("the energy law") expresses the fact that energy can neither be destroyed not created. The second law ("the entropy law") states that in an isolated system any spontaneous change leads to an increase of its entropy (see Chap. 7). Yet the knowledge that we associate with the laws of thermodynamics is not the basis of action in the sense of technological capability. On the contrary, the laws of thermodynamics tell us precisely what we *cannot* do, that is: we cannot construct a perpetual motion machine of the first or second kind. In other words: we cannot build a machine to produce energy out of nothing, or a machine to extract energy from some reservoir (such as the ocean) and turn it continuously into mechanical work. These examples illustrate the dangers entailed in general by regarding scientific knowledge to be as technological know-how as it is done by methodological constructivism.

4.3   Information in Living Matter

A philosophy of science which seeks its place beyond a mere feuilleton style of telling scienctists what sciene is has above all to face up to the concrete problems of science. It must show its ability to shed light where scientific correlations are unclear, where important problems obstinately resist solution, where deficiencies and anomalies of explanation appear or where the relevance of scientific statements is questionable. If this standard is applied to the philosophy of science developed by Janich, then with the best will in the world one cannot see where it has made any substantial contribution to the critical self-understanding of natural sciences. So far, this philosophy has amounted to nothing more than a rejection of everything, which smacks of naturalism.

Paradigmatic for this are the objections against the application of the concept of information in the natural sciences. Let us start by summarising the major points of the critique by following again along the line of the arguments, presented by Janich [7, 9]: Science, he argues, is in a state of hopeless confusion about the use of the concept of information. An almost limitless number of different disciplines has generated their own individual definitions and interpretations. Even more serious than the absence of an explicitly determined concept of information, however, would (according to Janich) be the notion entertained by some scientists that information must be regarded as a natural entity that is subject to natural laws. Repeating his arguments against naturalism, Janich insists that information is not a *natural*, but rather a *cultural* entity, the origin of which lies solely in the realm of communicative action of human beings.

As an example of the putative conceptual confusion in science, Janich refers to the use of the concept of information in biology. The idea of genetic information, he argues, is falsely seen as an objectifiable, material property of living systems. However, this view would invert the factual historical order in which the theory of information arose, i.e., beginning with human action and ending in the application to natural processes.

Thus Janich raises two fundamental objections. The first is related to the "undefined multiplicity" of concepts of information, the other to the "naturalistic interpretation" of information. Let us first look at the multiplicity of concepts. Here Janich makes a number of assertions that concern both the process of concept formation in science in general and the emergence of the scientific idea of information in particular. He interprets the allegedly "hopeless confusion" of the various information concepts quite generally as a symptom of the weakness of a science that is sapped of its vitality by adhering to its analytical self-understanding. Still more: Janich [7, p. 139, f.] makes the presumptuous statement that scientists deliberately avoid a strict definition of information in order to be able to present the lack of such a definition as plasticity and many-faceted applicability and thus as a positive aspect of the information theory.

We have to concede from the start that the concept of information is indeed extremely heterogeneous in science. "Information" is referred to both in the natural sciences and in the humanities, in numerous variants. We encounter it in physics, chemistry and biology just as we do in computer science, linguistics, cognition theory and the social sciences. Because of this broad area of its use, we may scarcely expect to find a single, unified definition of the term "information". By taking the example of biology we shall see that there are even good reasons for using completely different concepts of information, even within a single discipline.

When Janich asserts that the lack of a unified and explicitly defined concept of information point up a chronic weakness and indeed a wrong understanding of science, he completely ignores the way in which "everyday" scientific research is pursued. When a new theory is promulgated, then one always starts out from ideas that are initially blurred, ones that are more intuitively than rationally understood. The relevant concepts are then gradually sharpened up, in the course of a long process of interaction and exchange between the formation of new scientific terms and theories. Refined theories make precise scientific terms possible, while precise terms increase the power of theories.

The reciprocal dependence of the formation of terms and theories is well illustrated by the modern understanding of the origin of life. In the early years of research, numerous terms such as "organisation", "order", "complexity", "functionality" and the like were introduced into the theoretical approach to living matter, without any of these ideas being definable in any kind of precise way. Even the most basic term, namely that of "life" itself, could initially only be used in an imprecise and incomplete manner. Not until the development of the theory of the origin of life had reached a certain degree of maturity did it become clear that the incompleteness of the concept of life cannot be remedied—a logically inevitable consequence of the fact that the transition from non-living to living matter is a gradual one (see Sects. 2.2 and 3.2). One of the consequences of this is that the concept of life includes normative determinants that depend upon the particular field of research in which living systems are being investigated. Thus, the normative element, on which the constructivist philosophy of science sets such store, is revealed here as the actual cause of the "plasticity" of a basic notion of science, namely that of the variety and unsharpness of the notion of life.

The manner in which scientific concepts are formed is obviously much more complex than is assumed by the constructivists to be the case. It is simply untrue when Janich asserts that the lack of definition, frequently encountered in science, reveals "openness of the entire theoretical structure for empirically necessitated revisions, making a virtue of necessity" [7, p. 139]. Rather, the opposite is true: imprecision in definition, when it is not (as we saw for the case of the concept of life) fundamental in nature, can only be rectified in the course of the development of theory, step by step.

The complex reciprocal influences of the formation of scientific terms and the formation of theories need not necessarily lead to a unification of the concepts. On the contrary: the plurality of theoretical approaches, and the specialisation and refinement of theories, are precisely the characteristics of a sophisticated under-standing of reality, one that is inevitably reflected in the variety and the diversity of concepts. When, as in the case of the concept of information, one and the same term describes various real facts, then this does not (as Janich believes) signify a sys-tematic weakness that calls for a new context of justification. Rather, the ambi-guities of which Janich complains can generally be traced back to the terminological revision that becomes necessary when a term has to be adjusted to fit a changed theoretical context, an adaptation that always takes place with a certain delay.

This can be observed with especial clarity for the development of the information concept in biology. The first use of the term "information" was in connection with the idea of a genetic code. First indications of this association can be found in the brief treatise "What is life?" that was published in 1944 by the well-known physicist Erwin Schrödinger [13]. In this book Schrödinger addressed the fundamental problems of biology from the perspective of modern physics. Not least among these problems, and one that at the time was completely baffling, was the question of how the phenomenon of self-perpetuation and self-reproduction of organisms could be explained physically.

Before the era of molecular biology, the growth of living organisms was fre-quently compared to that of a crystal. As is well known, a crystal grows up when a microscopically small crystal seed, with a regular lattice structure, forms; further atoms or molecules then join this spontaneously. In this manner the structure of the elementary crystal is ultimately reflected in the structure of the visible crystal. This picture, in which microscopic order gives rise to macroscopic order, Schrödinger obviously had in mind when he expressed the hypothesis that the chromosomes might instruct the construction of a living organism in a way similar to that in which the microstructure of a seed crystal impresses its form upon the visible crystal. On the other hand, Schrödinger was aware of the fact that the immense complexity and diversity of living matter cannot be generated by the "dull" device of repetition [13, p. 4f.]. Therefore, he concluded, the chromosomes, as carriers of hereditary information, cannot have the simple order of a "periodic" crystal. Instead, he suggested that the chromosomes must be regarded as "aperiodic" crystals. Only in this way, he believed, could the ordered diversity of living matter become conceivable.

The chromosomes, Schrödinger elaborated on his idea, "contain in some kind of code script the entire pattern of the individual's future development and of its functioning in the mature state" [13 p. 21]. In the same context, Schrödinger states: "Let me use the word 'pattern' of an organism in the sense in which the biologist calls it 'the four-dimensional pattern', meaning not only the structure and functioning of that organism in the adult, or in any other particular stage, but the whole of its ontogenetic development from the fertilized egg the cell to the stage of maturity, when the organism begins to reproduce itself."[13, p. 20f.].

Thus, it is the process leading to the form and pattern of living matter to which the idea of the genetic code, and along with it that of information, was fitted. When, in this context, Schrödinger compared the genetic code to the Morse code, this was only to demonstrate the "variety of content compressed in the miniature code"[13, p. 61]. He was in no way trying to suggest any communication-theoretical parallels between biological transmission of information and the information theory developed later by Shannon. Rather, at the centre of his interest was the problem of coding in direct relation to the problem of material instruction, which raises the question of how a complex macrostructure can be contained in a simple microstructure (the chromosome).

What Schrödinger, with the state of knowledge of his time, described in a rather cumbersome, but valid, way was the fact that material complexity is related to the aperiodicity—that is, the incompressibility—of material structures and that the entire richness of living matter is deposited in this. The terms "aperiodicity" and "code" were used by Schrödinger more or less interchangeably, to describe the ordered diversity of living systems. Only later, after these terms had been made much more precise within the context of greatly differing theoretical concepts, did they come to differ in meaning.

The term "genetic code", for example, could only become more precise after the molecular structure of DNA had been elucidated by James Watson and Francis Crick. Only as a consequence of this discovery did it become known that living organisms contain a chemically based system of rules fixing a correspondence between the microstructure of the gene and the macrostructure of the organism, which is comparable to the coding structures of man-made communications systems. At the same time, this step paved the way towards the application of

Shannon-type information theory in biology. The term "aperiodicity", in contrast, was only much later given a precise meaning, within the framework of algorithmic information theory (see Sect. 3.2). Here it was found that aperiodicity, as envisaged by Schrödinger, is indeed a basic feature of complexity. In the case of biological information-carriers, the aperiodicity is expressed in the irreducibility of the sequences of genetic symbols.

The first insights into the information-theoretical foundations of living systems were thus obtained completely independently of any communication-theoretical problems, and certainly of any communicative actions of human beings. Only later did it become clear that there indeed is a strict analogy between the principles of coding in natural and man-made systems. As a consequence of this, certain developments in molecular biology and in communications technology, ones that to begin with took place independently of one another, later supplied one another with valuable ideas. The constructivistic mantra that the concept of information in the natural sciences refers back to Shannon's information theory, which in turn has its origin in human actions and interpersonal communications, is seen on closer inspection to be untrue.

4.4   On the True Nature of Information

The erroneous belief that the scientific concept of information originated in the non-scientific world of interpersonal communication and mutual understanding may be a consequence of the fact that Shannon originally termed his approach "communication theory" [14]. However, he wanted to avoid any explicit allusion to the various forms of interpersonal communication. On the contrary, as a signals engineer Shannon was only interested in improving the conditions and possibilities for an error-free transmission of information. In contradistinction to human communication, this has nothing to do with the actual content of the information that is being communicated.

Shannon's theory is thus only concerned with the machinery of sending and receiving messages and the channels through which these pass. As we have seen, this restriction is already expressed by the statistical definition of information, in which only sequences of abstract symbols are considered. The Shannon information also reveals the true character of the concept of information, and at the same time it provides us with an answer to the question of whether information is in fact a natural entity.

For this purpose we again refer to the entropy H of a source of messages (Sect. 4.1). It has already been mentioned that the entropy of a message source corresponds mathematically to the entropy as defined in statistical physics (except for its sign). In precise terms: we are dealing here with two distribution functions, of which one describes a distribution of messages and the other a distribution of energy over various quantum states. According to this structural analogy, information can be equated with "negentropy" [3].

The first scientist to point out that the entropy function is at the same time a measure of information was Ludwig Boltzmann [2], the pioneer of the statistical foundation of entropy. For the case of an ideal gas, this analogy of entropy and information is immediately comprehensible: As is well known, the temperature of an ideal gas is directly proportional to the average kinetic energy of the gas molecules. Thus, the temperature is an *intensive* quantity. Its value does not depend upon the actual size of the system, which is given by the number of gas molecules that go into the calculation of the average. In consequence of this finding, we are forced to conclude that there must be another quantity, complementary to temperature, that is *extensive* in character—one that does depend on the extent or quantity of gas present, otherwise our knowledge of the total thermal energy of the system would be incomplete. This extensive quantity is precisely the entropy.

The entropy function thus provides us, as Boltzmann recognised, with a measure of the loss of information that occurs in statistical physics when one forgoes exact knowledge of the microstate. By doing so, Boltzmann was also the first to set out in detail the dependence of scientific knowledge upon the means used to describe, and to express abstractly, the system being considered, although he did this in a manner completely different from that demanded of science by the constructivists.

Nevertheless, Janich saw in the information-theoretical interpretation of entropy by Boltzmann a direct confirmation of his own thesis, according to which the concept of information in physics has—"via mathematical statistics"—acquired a non-permissible, naturalistic extension, so that today "information [is regarded] as a structural element on an equal footing with matter and energy" [7, p. 145]. However, by using the term "structural element" Janich has, even if unintentionally, described precisely the actual character of information. This is because equating information and negentropy gives the information concept not a naturalistic, but a structuralistic interpretation. At that point the propagated accusation that science falsely regards information as a natural entity implodes completely.

But: What, after all, is a natural entity? Reading Janich's exposition of this question, we gain the impression that the idea of a natural entity is restricted to material objects and their intrinsic properties. Indeed, such a view of the essence of a natural entity is problematical, as it severely restricts the area of validity of scientific statements. The natural sciences are apparently only in a few cases concerned with the primary properties of Nature. Rather, they principally address the relationships between natural objects—for example, the interactions between, and distributions of, material and energy states, which are decisive for any causal-analytical understanding of Nature. If, however, the concept of a natural entity is restricted a priori to material objects, then it is scarcely possible to speak meaningfully of natural phenomena.

The example of entropy in physics makes this very clear. Entropy is certainly an object of scientific knowledge, one that can be described by a physical quantity; it is however not a natural object in the way that trees, stones or other material things are. The entropy of a system is a statistical distribution function; its character as an "object" consists solely in the object-like character of the energy distribution that it describes. When Janich asserts that the adoption of the concept of entropy in

Shannon's information theory and its transference back to the information-theoretical usage of the term entropy in physics had ultimately led to the naturalistic understanding of information, his assertion has to be qualified by bearing in mind that equating information with negentropy is only a statement of the structural analogy of two distribution functions. Indeed, Shannon only wanted to emphasise the formal analogy between informational entropy and physical entropy, when—on the advice of John von Neumann—he used the letter H to denote it (thereby drawing a parallel with Boltzmann's H function).

Thus, information is not a natural entity, but merely a concept that corresponds structurally to that of entropy. This implies that information theory does not belong either to the natural sciences or to the humanities. Rather, it must be allocated to the rapidly expanding group of so-called "structural sciences", which alongside information theory includes such important disciplines as game theory, cybernetics, systems theory, complexity theory and others (Chap. 9). The structural sciences deal with the abstract structures of reality. These are investigated independently of the question of where these structures are encountered in the real world—in living or non-living Nature, in artificial, natural, simple or complex systems.

Let us finally take an example that leads immediately to the concept of information. It is the category of form. This category is basic within science insofar as there is no "matter without form". But "form without matter" is very well imaginable. More than that, "form" is a fundamental category of human perception. It is an abstract term and, as such, it is basic and typical for the structural sciences and their goals of perception.

In fact, the concept of form is closely related to that of information as used in classical information theory. Thus, for example, the "form content" of any structure (that is, the measure of its form) can in general be conceived of as the number of decidable alternatives (yes/no decisions) associated with this structure. Similar considerations lie at the bottom of Carl Friedrich von Weizsäcker's [16] reconstruction of abstract quantum theory. This approach to the form content of objects is also encountered in the selection of a message from a message source. It is identical to the information measure introduced by Hartley and Shannon.

Moreover, the form content, as well as the information content, depends upon the questions being asked and, thereby, upon the particular perspective one adopts [10]. For example, the form content of a DNA molecule is not the same for a physicist as it is for a molecular biologist. The physicist would presumably be interested in the highly complex structure of the molecule and would try to find out all its details, perhaps by X-ray analysis. For the molecular biologist, as a rule only the sequence of the building-blocks of the DNA would be of interest. He would start by trying to determine this sequence by chemical analysis. In the one case, the form content would depend upon the positions in space of all the atoms in the molecule; in the other it would depend upon the length of the nucleotide chain.

One and the same natural object can thus, depending upon the context in which it is being considered or investigated, have different form content and thus contain different structural information. This shows that information cannot in itself be a natural entity. Rather, information is primarily a mere abstraction, which only takes on a concrete nature when it becomes associated with a natural object that confers this form upon it.

Probably the most impressive example of the abstract character of information is to be found in mathematical physics. There, natural laws are frequently expressed in the form of differential equations, such as can only be solved unambiguously when certain initial and boundary conditions are given. The latter have the function of selection conditions, ones that restrict the set of possible processes, allowed by natural law, to those that actually take place in the real world. For this reason, the natural laws always have to be supplemented by initial and boundary conditions if we are to make any concrete statement about the world and the events taking place within it. In other words, the initial and boundary conditions express facts about the world. Only through these conditions do the natural laws attain a relationship with the event character of the real world. One might also say that the initial and boundary conditions operate through the natural laws to impose structure upon the world. In this sense they are at the same time carriers of information (see also Chap. 8).

Everything that we discussed above referred to the question of the true nature of information. The analysis has clarified once more the fact that information is a key concept for the understanding of living matter. The logical next question of whether beside the structural or syntactic aspect of information its semantic and pragmatic dimension can also become the subject of the exaxt sciences will be looked at in detail in the next chapter.

References

1. Albert, H.: Treatise on Critical Reason. Princeton University Press, Princeton (1985). [Original: Traktat über kritische Vernunft, 1980]
2. Boltzmann, L.: Lectures on Gas Theory. University of California Press, Berkeley (1964). [Original: Vorlesungen über Gastheorie, 1896/1898]
3. Brillouin, L.: Science and Information Theory. Academic Press, New York (1956)
4. Eibl-Eibesfeldt, I.: Human Ethology. Aldine De Gruyter, New York (1989). [Original: Die Biologie des menschlichen Verhaltens, 1984]
5. Hartley, R.V.L.: Transmission of information. Bell Syst. Tech. J. 7(3), 535–563 (1928)
6. Janich, P.: Naturgeschichte und Naturgesetz. In: Schwemmer, O. (ed.): Über Natur, pp. 105–122. Klostermann, Frankfurt am Main (1987)
7. Janich, P.: Grenzen der Naturwissenschaft. Beck, München (1992)
8. Janich, P.: Erkennen als Handeln. Von der konstruktiven Wissenschaftstheorie zur Erkenntnistheorie. Jenaer Philosophische Vorträge und Studien Nr. 3. Palm & Enke, Erlangen/Jena (1993)
9. Janich, P.: Information als Konstruktion. In: Max, I., Stelzner, W. (eds.): Logik und Mathematik, pp. 470–483. Springer, New York (1995)
10. Küppers, B.-O.: Information and the Origin of Life. MIT Press, Cambridge/Mass. (1990). [Original: Der Ursprung biologischer Information, 1986]
11. Lorenz, K.: Behind the Mirror. A Search for a Natural Historyof Human Knowledge. Harcourt Brace Jovanovich, New York (1977). [Original: Die Rückseite des Spiegels, 1973]
12. Rényi, A.: Probability Theory. North-Holland Publishing Company, Amsterdam (1970)

13. Schrödinger, E.: What is Life? (Combined reprint of „What is Life, 1944" and „Mind and Matter, 1958"). Cambridge University Press, Cambridge (1967)
14. Shannon, C., Weaver, W.: The Mathematical Theory of Communication. University of Illinois Press, Illinois (1964)
15. Weizsäcker, C. F. von: Unity of Nature. New York (1980). [Original: Einheit der Natur, 1971]
16. Weizsäcker, C. F. von: The Structure of Physics. Springer, Heidelberg (2006). [Original: Aufbau der Physik, 1985]

Chapter 5
Is Language a General Principle of Nature?

Communication takes place not only among humans, but also among animals, plants and bacteria. Even the molecules of heredity, as carriers of genetic information, appear to be organised according to the principles of language. Is there really a "language" of molecules, underlying every living thing? Is language a universal principle of Nature?

© Springer International Publishing AG 2018
B.-O. Küppers, *The Computability of the World*, The Frontiers Collection,
https://doi.org/10.1007/978-3-319-67369-1_5

5.1   The Structure of Language

Information and communication are regulatory principles which are indispensable
for the preservation of the functional order in living matter. The intricate networks
of the processes that constitute life functions would collapse if they were not
permanently stabilised by information. This in turn calls for a perpetual exchange of
information at every organisational level of living matter. In other word: all
essential life phenomena require what we usually term "communication".

Already here we run into a fundamental problem. Communication presupposes
the existence of some kind of language in the broadest sense. Is it—under this
precondition—conceivable that the material carriers of life functions, such as biological macromolecules, communicate one with another, in order to maintain the
complex order of living matter? Does any language exist at all outside the realm of
human language? Put differently: Is language a general principle of Nature, one that
manifests itself not only among humans, but also among animals, plants,
micro-organisms and even molecules?

Johann Wolfgang von Goethe, who observed Nature with the utmost meticu-
lousness, seems to have regarded this idea as definitely conceivable. In the foreword
to his treatise on the "Theory of Colours", he writes enthusiastically about the
language of Nature: "Thus, Nature speaks down towards other senses—to known,
poorly known and unknown senses; she speaks to herself and to us, through a
thousand appearances" [12, p. 315]. At the end he summarises: "So multifarious, so
complex and so incomplete this language may sometimes appear to us, its elements
are always the same" [12, p. 316].

Goethe is obviously not just referring to the language of Nature in some poetic
way. He believed that natural phenomena genuinely reflect the elements of a universal language. This leads on to the question of whether one can really develop a
consistent picture of language that on the one hand suffices to describe human
language and on the other hand is so abstract that it can be applied to the "thousand
appearances" of Nature.

In fact, it is a recognised finding of comparative behavioural research that
primitive forms of communication exist in the animal kingdom (see [3]). Plants,
too, can exchange information with one another by using a sophisticated communication system that employs certain scent molecules; in this way they can defend
themselves against impending attack by pests (see [15]). We have long known that
bacteria communicate by releasing chemical signals. Maybe they even use some
kind of a linguistic code. But most surprising, perhaps, is the fact that basic
organisational structures of human language are already reflected in the molecular
structure of the carriers of genetic information. This analogy goes far beyond a mere
vague correspondence—it comprises several almost identical properties that are,
without exception, found in all forms of living beings (see [9, 16]).

If, in the light of this, we set out to uncover the "language of Nature", then we must first ask some questions about the "nature of language". This will, repeatedly and inevitably, lead us back to the language of humans, as this is paradigmatic for our understanding of language. However, we cannot simply extrapolate in an abstract way from human language. In trying to do so we would lose sight of the unique position of humans as beings that *understand* and thus also of some unique properties of human language, in which we amongst others pronounce judgements, formulate truths, express opinions, convictions and wishes. Moreover, at the level of interpersonal communication there are various different ways in which language appears: it may employ pictures, gestures, sounds or written symbols. Therefore, to develop a general concept of language, we have to construct a notion of language from the bottom up. We must try, on the basis of highly general considerations, to develop an abstract concept of language that on the one hand reflects human language and on the other hand is free from its complex and specific features.

This proviso places first of all a "structural" aspect of language in the foreground that is related to all forms of communication. First of all, it is evident that successful communication is only possible when the communicating partners use the same inventory of signs. In addition, the sequences of signs have to be structured according to the same rules and principles. The unified syntactic structures then make up a language that is common to the sender and the recipient, even though this is still only a pure "sign language".

However, the syntactic structures of a language must still fulfil another condition in order for them to take on their function as elements of a language: the sequences of signs on which the language is built up must be more or less aperiodic. Only in aperiodic sequences can sufficiently complex information be encoded. This is demonstrated by human language, whose richness of expression is to be traced to the aperiodicity of its syntactic structures.

Let us take another look at the basics of language. Its character can most clearly be illustrated by considering the communication processes that take place in a living organism. Language here has primarily *instructive* properties, in that it—on the basis of genetic information—sets the milestones which the innumerable elementary processes in the organism have to follow.

Nevertheless, the dynamics of a living system by no means represent a clockwork-like mechanism, confined to following a rigid set of instructions. Rather, the instructions encoded in the genome change perpetually during the course of expression of the genetic material. This is due to the fact that the physical and chemical conditions under which the genome becomes operational in the living cell are inevitably modified themselves with each step of expression of the genetic information. Thus, as the information content of the genome derives its meaning only in the context of its cellular environment, the information content changes in line with the change of its physical and chemical boundaries. As a consequence of the indissoluble interaction between the genome and its environment, a living system represents an extremely complex network of feedback loops which needs perpetually to adapt and reorganize its internal statics and dynamics in the course of the progressing expression of the genome. To do this, a living system must continually take up, evaluate and process fresh information from the current state of its inner organisation and its outer environment. This vigorous interplay and exchange of information, and its continual re–assessment, transcends mere instruction, since at this level of self-organisation certain order principles are required, which define the rules of communication and which correspond to what we denote as language.

To appreciate the full breadth of this, we take another look at human language: its high degree of context-dependence. We are not always consciously aware of this fact in our daily use of language. But the context-dependence of language immediately becomes clear when we think of ambiguous words or phrases. For example, the expression "dumb waiter" might mean a waiter who is incapable of speech, or it might mean a lift for conveying food from the kitchen to the dining-room. To decide which meaning is intended, one needs to refer to the overall context in which the phrase is used.

Taking examples from everyday language the philosopher Günther Patzig [23] has illustrated that the act of speaking is always performed in a context of other actions. Thus, the curious word sequence "right, left, head raised, lowered, side, good, now rear left, right, brakes, fine" will only be comprehensible for someone who has taken a car into have its lights tested. Without a context, garbled word sequences of that kind convey no meaning and are incomprehensible. Even fragments of speech such as "Two to Berlin and back" or "Sausage, sir?" are examples of spoken expression the meaning of which is immediately clear when the context of the action is already known: the one at a ticket window, the other in a restaurant.

The context-dependence of spoken phrases is a universal characteristic of language and is reflected even at the deepest level of human language. This is because sounds, words and the like stand in a highly complex fabric of relationships one to another. More than that: this network of relationships is what makes them elements of language at all.

The "inner" structure of human language was first studied systematically by Ferdinand de Saussure [6] at the beginning of the 20th century. On the basis of his research, a linguistic-philosophical school of thought emerged at various European centres—above all in Prague, Moscow and Paris—which later became known as "structuralism".

Structuralism presents the view that every language is a unique network of relationships between sounds, words and their meanings—one that cannot be reduced to its basic components, because each component only receives a meaning within the context of the overall structure. Thus, only through the overall structure do the elements of the language coalesce into a linguistic system. This system provides a framework in which the linguistic elements are demarcated over against each other, and a set of rules is assigned to each element.

There are numerous examples to illustrate how the elements of a language depend upon the overall structure of language (see for example [22, p. 101 ff.].

Thus, in English the sounds "L" and "R" differ because they have a function to differentiate the meanings within word-pairs such as "lack" and "rack" or "loot" and "root". However, the situation in Japanese is different: the "L" sound and the "R" sound do not differ, because in the context of the Japanese language they have no function in differentiating between meanings. A somewhat similar example is familiar to speakers of English: here, there is a clearly audible difference between the alveolar "R" and the rolled "R", but in English this is not associated with any difference in meaning; at most, the difference indicates which linguistic region the speaker hails from.

What applies to sounds applies also to words [22, p. 105 f.]. The internal reference of a word to the structure of the language in which it is used is also the reason why word-for-word translations are generally not possible. Thus, the English word "horse", the German word "Pferd", the Latin word "equus" and the Greek "hippos" all refer to one and the same biological species, but their linguistic meaning first crystallises out in their demarcation over against alternative terms and, furthermore, depends upon the corresponding linguistic structure. For example, in English the meaning of the word "horse" becomes more precise in the context of alternative expressions such as "stallion", "mare", "gelding", "colt", "foal", "roan", "dapple", "sorrel", "steed" or "nag".

Each one of the many human languages has its own structure (see [14]). However, this is demonstrably only true of the so-called surface structure of a language. In respect of the deep structures of language there may indeed be general principles. At any rate, this appears to be the case, as shown in studies by the linguist Noam Chomsky. Briefly stated, Chomsky [5] claims that there is a "universal grammar" that is common to all human languages and that it lays down uniform rules for their constitution. This basic grammar, according to Chomsky, is an inborn—that is, genetically hard-wired—linguistic structure common to all humans, and among other things it enables children to learn complex languages in a very short time. Even if Chomsky's ideas are controversial, they do point a way to how universal principles of language may possess a deep-seatedness that reaches down even to the genetic level.

Nonetheless, Chomsky's theses are based upon human language. Yet the question with which we began is much more far-reaching: we wanted to know whether language is a general principle of Nature. In the light of our discussions so far, this question can now be put in a different way: Can the structuralistic view of language be reversed and every ordered network of structural elements be regarded as linguistic structure? Such a radical reverse is by no means new; it has for a long time been at the centre of a powerful philosophical school of thought that has its origin in France and leans heavily on linguistic structuralism: one speaks of the so-called "French structuralism".

Its central message has been expressed by the philosopher Gilles Deleuze in the following way: "In fact, language is the only thing that can properly be said to have structure, be it an esoteric or even non-verbal language. There is a structure of the unconscious only to the extent that the unconscious speaks and is language. There is a structure of bodies only to the extent that bodies are supposed to speak with a language which is one of the symptoms. Even things possess a structure only in so far as they maintain a silent discourse, which is the language of signs" [7, p. 170 f.].

The ethnologist Claude Lévi-Strauss took up this thought and applied it consistently in his anthropological studies. He attempted to interpret the rules of human conduct—up to the structure of social institutions—as structures of an over-arching social language. Other disciplines too, from linguistics through literary criticism, sociology, psychoanalysis to mathematics, all showed a strong interest in the abstract structures of reality, and thus drew major intellectuals such as Roman Jakobson, Roland Barthes, Gilles Deleuze, Michel Foucault, Jacques Lacan and the Bourbaki group of mathematicians under their spell.

The breadth and depth with which structuralism contributes to our world-view may well be a matter for debate. A final judgement will depend upon whether structuralism is understood as a merely methodological, though relatively exact, tool of the humanities and social sciences, or whether it is understood as the final court of appeal for our understanding of the world. Insofar as structuralism recognises the boundaries that circumscribe every exact science, it will not claim to embrace reality in its entirety and with all its facets. Rather, it will adhere to its original programme and investigate the structures of reality as such—that is, independently of which forms these structures manifest themselves in.

5.2   Towards a Radical Linguistic Turn

The exchange of information serves the mutual understanding between sender and recipient. For this reason, the concept of communication always contains an echo of the concept of understanding. In this important point, the concept of communication differs from that of information. This difference can even be recognised in the etymological roots of "information" and "communication". The concept of information—completely in the sense of the Latin root *informare*—implies at first the formative and instructive functions of a signal, message etc. (see Chap. 4). The concept of communication (*communicare*: to perform collectively, to impart) emphasises rather the process whereby sender and recipient try to reach a mutual understanding.

In other words: The object of understanding is the content of the information communicated. The purpose of communication is to reach an accord in the assessment of the information that is passed from the sender to the recipient; in some cases this will also lead to joint action. In contrast, the mere instruction that is contained in a piece of information acts like a command, conveying some kind of directives from the sender to the recipient.

Since we wish to extend the concept of communication to include natural processes, we must explain the concept of mutual understanding in more detail. Intuitively, one tends to reserve the idea of communication for the exchange of information between human beings, because the idea of "mutual understanding" outside the sphere of human consciousness seems to be meaningless. However, that view should be examined critically as it could be a too narrow interpretation of the concept of understanding.

The primary purpose of communication is to enable the partners involved to reach an agreement on some issue. To achieve this, they must of course make it clear to one another that one partner is going to inform the other. However, mutual understanding does not necessarily presuppose any reflection on the issue to be communicated. Rather, it may already exhaust itself in the mere exchange of information, which may induce some co-ordinated actions without asking what the meaning of these actions is. The mutual understanding of the meaning of information reduces here to the practical or pragmatic implementation of the information being communicated.

The concept of "common understanding" can obviously be understood in different ways, with different scopes. If we are prepared to abstract from the manifold forms of human understanding and to reduce the concept of understanding to the mere fact of reaching a consensus with respect to common actions, then we can easily see how this principle can operate in living beings at all levels. In that sense molecules, cells, bacteria, plants and animals all have the ability to communicate. Here, communication is nothing more than the co-ordination and fine-tuning of processes by means of chemical, acoustic and optical signals.

With this argument, we again tread a path that some philosophers of science, in a somewhat contemptuous tone, criticise as "naïve" naturalism. This criticism is primarily directed at the idea that information exists outside the various forms of human communication, i.e., that information exists as a natural entity (see Chap. 4). It is based upon the conviction that only human language can carry information and, in consequence, it rejects completely the application of linguistic categories to natural phenomena. For these critics, expressions such as "information" and "communication"—common parlance in modern biology—are nothing more than metaphorical figures of speech, and this usage bears witness to a highly uncritical attitude in the use of terms such as "language" and "understanding".

We wish here to examine this objection in more detail, and we do so by taking another look at the question of what exactly we are to understand by "understanding". In expressing this question tautologically we make it clear that the problem of understanding can quickly lure the unwary into a circular argument: Any philosophical doctrine that sets out to "understand" understanding is forced to presupposes the existence of a concept of understanding.

As a fundamental feature of any kind of understanding, we may state that one can only understand something if one has already understood something else. This elementary insight stands at the centre of philosophical hermeneutics, which is surely the most influential doctrine of understanding (see for example [9]). The assertion that any understanding is dependent upon another, prior understanding makes immediate sense. We experience it all the time in our daily communication with other people. This experience tells us that any successful and meaningful exchange of information can only take place on the basis of some advance information. The reason for this is that information in an absolute sense does not exist.

Information is always a relative entitity, because it receives its meaning only in relation to some other information.

Thus, the basic thesis of philosophical hermeneutics, according to which one can only understand something when one has already understood something else, seems to get to the heart of the problem of understanding. If we interpret this in a relativistic sense, it would fit to the character of information presented in this book. However, this way of viewing the issue would completely contradict the intentions of philosophical hermeneutics, which interpret its basic thesis in a completely different sense. According to that interpretation, every "advance" understanding, which is the basis for any other understanding, results from the "rootedness" of human beings in their life-world. Thus the validity and the claim to truth of human understanding are anchored in human existence, so that existence becomes an absolute norm for understanding.

However, the openness and the relativity of human understanding, essential as they are for "critical" thinking, are inevitably lost in this kind of philosophy. In fact, philosophical hermeneutics propagates a fundamentalism of understanding, one which fits into the tradition of absolute understanding of the world and not at all into a relativistic understanding of it as prompted by the modern world view of critical sciences.

We gain access to the world by means of language. Therefore, language is from the viewpoint of philosophical hermeneutics the gateway to the understanding of our existence. Hans-Georg Gadamer expressed the guiding principle of this philosophy in the frequently cited dictum: "Being that can be understood is language" [11, p. 470]. If this is true in a rigourous sense, if human language is the precondition for any understanding, then language becomes the final, unsurpassable reference point for access to our world.

However, the statement "Being that can be understood is language" could even be interpreted more radically, as done by the philosophers of French structuralism (see above). According to their interpretation, "being" can only be understood at all if it has the structure of language. Thus language becomes promoted to being a universal structure of our world, one without which we would have no way of recognising the world [18]. This idea would induce the most radical linguistic turn we can think of, namely the turn from the "structure of language" to the "language of structures" [19].

Such an interpretation of language would certainly turn the intentions of Gadamer upside down. From his point of view, *human* language is the ultimate access to our world and not an abstract language of structures. On this premise it is completely unacceptable when natural scientists speak of the "language of Nature". This, Gadamer would have argued, gives the impression that language is a natural principle, which exists independently of the human observer and manifests itself in the material structures of Nature. Scientists who speak and think in these terms seem to behave like wrong-way drivers on the motorway, trying to make sense of the world by pursuing linguistic arguments that run directly counter to the direction in which philosophical hermeneutics point us.

Is all talk of a "language of Nature" really mere rhetoric? An outside observer, watching these opposing camps from a distance, might indeed be somewhat surprised by this controversy, and might regard it as a bizarre dispute about ideas as it seems to be typical for philosophers. He would point out that we inevitably represent scientific knowledge by using human language, that in doing so we always take concepts out of their original context and transfer them to another context, and that—viewed in this way—any description of the properties of Nature has (more or less) a metaphorical character. On the other hand, it cannot be denied that concepts such as "information", "communication" and "language" have proved immensely valuable in describing the processes taking place in living organisms. In other words: do not the spectacular developments of modern biology justify the use of such concepts?

The massive resistance that the information-theoretical orientation of modern biology has encountered in the humanities is due above all to the fact that concepts such as "information", "communication" and "language" are always loaded with semantic notions like "content", "value", "truth", "understanding", "purposefulness" and so forth. This brings us to the actual nub of the controversy. Phenomena of meaning, as expressed in the semantic dimension of information, communication and language, appear to lie completely outside the framework of naturalistic explanations and, therefore, to be inaccessible for use by the exact sciences.

The philosophical controversy about this question is by no means centred only around disputes over the different use of concepts in science; rather it bundles, like a burning-glass, the deep divide between the natural sciences and the humanities. As the biological sciences stand at the boundary between these two great currents of science, it is no surprise that they have been the first to come into the line of fire. With the molecular foundation of biology, issues such as the intrinsic meaning of genetic information move into the focal plane of scientific explanation. Not least, this includes the wonderful plan-like and purpose-like behaviour of living Nature, which has always been among the great mysteries of the world because it appears to transcend the boundaries set for explanation by the exact sciences. Nevertheless, let us take up this mystery as a challenge and try to find out, how and to what extent this seemingly insoluble problem could be approached.

5.3   Approaching the Semantic Dimension of Language

There are a number of plausible reasons for the assumption that the semantic dimension of information is incapable of any explanation based upon natural law. First of all, meaningful information, as we have already seen, is always "assessed" information. However, the value of a piece of information is never an absolute quantity, but is always to be seen in relation to the recipient of the information. Therefore, the meaning of information depends fundamentally upon the current state of the recipient who is receiving the information. For example, in interpersonal communication this state is determined by the prior knowledge possessed by the recipient, and also upon the recipient's prejudice, desires, expectations, and so forth. In short: the valuation standard which a recipient applies to the content of some information is unique in the sense that it depends not only upon the particular pre-history of the recipient but also upon the particular context in which he receives the information.

This leads us on to a further question: Can the particular, as it is, ever become the object of study by a science that is based upon generalised concepts and law-like behaviour? This question is so fundamental that it was already investigated critically by Aristotle. For him, the logician, there could never be any general understanding of the particular because, already for logical reasons, the general and the particular were mutually exclusive—a view that has persisted up to our day and has set a decisive imprint upon our scientific understanding.

In this context, the efforts made by the cultural philosopher Ernst Cassirer [4] to resolve this contradiction seem particularly valuable. Cassirer, swimming against the Aristotelian tide, endeavoured to bridge the presumed antithesis between the general and the particular. The idea that such a bridge might be possible had already appeared in an *aperçu* of Goethe: "The general and the particular concur: the particular is the general, appearing under different kinds of conditions" [13, p. 302]. Precisely this thought also underlies the approach to the particular, as developed by Cassirer. The particular, he argued, does not become particular because it eludes all general determinations. On the contrary, it becomes the particular, because it enters into more and more specific relationships with the general. It is thus ultimately the unique network of relationships to the general that allow the particular to appear as such. This is a highly interesting figure of thought, as it points a way to how one might understand the particular in a stepwise manner through the general, although it can never can be generalizied in an exhaustive manner (see the detailed discussion in Sect. 9.2).

On the basis of this idea we will now develop a scientific approach to the semantic dimension of information that even allows the formulation of a measure of meaning [19]. To do this, we must first look at the general aspects that—in the sense explained above—are constitutive for the semantics of information. Without attempting to be comprehensive, we can immediately mention three aspects that can be associated with the notion of the meaning of information: the novelty value, the pragmatic relevance and the complexity of information. Of course, we could think of many other aspects that may be constitutive for the assessment of the content of some information. However, our choice serves only the purpose of demonstrating the general idea according to which semantic information can be objectified. In contrast to other attempts to measure semantic information, this approach is not based on a special aspect of semantics, but it combines the main aspects discussed in the past in a completely novel manner

Let us first consider these aspects separately. Already in the 1950s, the logicians Yehoshua Bar-Hillel and Rudolf Carnap developed, on the basis of linguistic studies, a concept that was intended to allow the content of information to be quantified [1]. They took up the idea of the information engineer Claude Shannon,

according to which the essential property of information is to eliminate uncertainty
(see Sect. 4.1). Setting out from this idea, they projected Shannon's measure of
information onto statements in language and measured the information content
according to the criterion of how strongly the range of expectation of all possible
linguistic statements was restricted by the linguistic information in question.
Admittedly, this concept that reduces the semantics of information to its novelty
value could only be applied to an artificial language. For this reason the range of
applicability of this approach has remained extremely narrow.

In comparison with this, the idea of measuring semantic information by its
pragmatic relevance has been much more fruitful. This approach, developed by
Donald McKay and others at the end of the 1960s, takes account of the fact that
information rich in content may provoke a reaction on the part of the recipient [21].
Such reactions can in turn be expressed in concrete actions and can thus be used as
measurable indicators of the content of the information in question. The "pragmatic
relevance" of a piece of information is however in no way tied down to its having a
human recipient who acts consciously. Under certain circumstances, the pragmatic
relevance of a piece of information might only be observable within the functional
structures of the receiving system, as for example is the case in a living cell when
the cell is reacting to a certain stimulus. In fact, the pragmatic or functional aspect
of semantic information has been investigated in particular depth in the context of
biological systems. Thus, it could be demonstrated that the evaluation of the
semantics of genetic information takes place through natural selection [8]. In this
case, the pragmatic aspect of genetic information could be even objectified and
quantified by its selection value.

Another theoretical approach, alongside the aspect of selectivity, is aimed at the
conditions that the syntax of information must fulfil if it is to be a carrier of
semantics at all. As we have already seen, one such condition is the aperiodicity of
a sequence of symbols, as only aperiodic sequences offer a sufficiently wide space
for coding meaningful information. It is the same idea by which Erwin Schrödinger
in the 1940s was guided in developing his concept of the chromosomes as aperiodic
crystals (see Sect. 4.3). This idea also promotes the aperiodicity of the syntax of a
symbol sequence to being a meaningful measure of the complexity of the semantics
built up over it, and thus to a general characteristic of semantics [17].

At this point one might perhaps have gained the impression that there is some
inconsistency among these very various aspects of semantic information. However,
that is not the case. On the contrary, the breadth of this variation is a necessary
consequence of the idea, described above, according to which the characteristics of
semantic information arise through the manifold relationships to its general determinants. Such determinants may be considered at it were as "elements of a semantic
code" [19]. However, in contrast to its traditional usage, the term "code" does here
not refer to a set of rules for assignment or translation of symbols, but rather to a
reservoir of value elements, from which the recipient configures a value scale for
the evaluation of the content of the information received. Strictly speaking, the
semantic code is an instrument of evaluation which, by superimposition and
specific weighting of its elements, restricts the value that the information has for the

recipient. In this way it becomes a measure for the meaning of information. The information value is a relative and subjective measure insofar as it depends upon the evaluation criteria of the recipient. However, for all recipients who use the same elements of the semantic code, and who for a given message assign the same weights to the elements, the information value is an objective and over-arching quantity.

As mentioned above, typical elements of the semantic code are "novelty", "pragmatic relevance" and "complexity". However, these elements are far from exhausting the semantic aspect of information. Only when the general aspects of a piece of meaningful information have been disclosed from all possible perspectives does its semantic dimension finally appear in all its substantial richness. In other words: The more elements are included in the evaluation scale of the recipient, the finer will be the semantic structure that emerges.

However, even single elements of the semantic code already allow deep insights into the character of semantic information. Let us consider for example the element of "complexity". The enormous breadth of conclusions that this aspect of semantic information makes possible will be illustrated by using an example from human language. To this end, we again take up the central thesis of philosophical hermeneutics, according to which one can only understand something when one has already understood something else. We shall first express this thesis in a quantitative manner. In the parlance of information theory this thesis invokes the question: How much prior information is needed for a given piece of meaningful information to be understood?

This question would at first seem to evade an exact answer, simply because it—once more—involves the problematical concept of understanding. Surprisingly, however, that an answer is entirely possible. But to do this we must restrict ourselves to the minimum condition for understanding. This basic condition obviously consists in first registering the information that is to be understood. Provided that the incoming information has a meaningful content the sequence of symbols must have a (more or less) aperiodic structure that is equivalent to a high complexity (see also Sect. 3.2). For sequences of that kind, however, no algorithms exist that would allow knowledge of a part of the sequence to be used to predict the rest of it. The recipient must therefore have registered the entire sequence of symbols that embodies the information in question, before the actual process of understanding can begin. More precisely: The mere act of registering a meaningful item of information thus requires a quantity of information that has at least the same degree of complexity as the information that is to be understood.

Let us illustrate this result again with an example from human language. For this purpose we consider a sentence that has a meaning for us:

THE WORLD IS TRANSPARENT AND COMPUTABLE

We are not interested in whether this statement is true. We are solely concerned with the complexity of the sequence of letters, which we regard in turn as a measure of the complexity of its information content. In this case the complexity of the letter sequence is as great as it could be, as there is no algorithm that is shorter than the sequence and in which at the same time the complete sequence is encapsulated. In this example, it is obvious that the sequence of letters is aperiodic and, in that sense, random. If the sequence had been periodic, or largely periodic, then because of its internal regularity—or law-like structure—it would have been possible to compress it, or, conversely, to generate it from a shorter algorithm.

Thus, viewed from a syntactic standpoint, meaningful sequences of letters are always random sequences. However, care should be taken: the converse of this conclusion is not true. Not every random sequence of letters is a meaningful one. These results are fundamental. They remain true even when we take note of the fact that in every language there are syntactic rules according to which the words of that language are put together to form correct sentences. However, these rules only serve to restrict the size of the set of random sequences that can make sense at all. They do not allow us to draw any conclusions about the actual content of a sentence, that is, its meaning; the latter always depends upon the unique sequence of letters of the sentence concerned.

We can put this in another way: Information that bears meaning cannot be compressed without changing its meaning. It may indeed be possible to press the information content into its bare essentials, such as telegram or headline style, but even then some information gets lost.

Nevertheless, it cannot be rigorously proved that a compact algorithm does not exist for the above sequence. Perhaps there might be some simple rule, concealed in the sequence that we have not detected so far. However, that is arbitrarily unlikely, as almost all binary sequences (as we demonstrated in Sect. 3.2) are of the greatest possible complexity and are therefore aperiodic or random. In human language, this fact is reflected in the so-termed "arbitrariness" of semantics. This is a technical term in linguistics: it expresses the obvious issue that there is no regular relationship, no law-like connection, between syntax and semantics. Consequently, the mere analysis of the syntax of a sentence does not allow one to derive the corresponding semantics.

Let us summarise these results. To understand a piece of information, one always needs background information of at least the same complexity as the information that is to be understood. This is the answer that we sought to the question of how much information is needed to understand some other information. At the same time, we have provided the phenomenon of the context-dependence of information and language a highly precise form.

Yet, from this a number of questions arise: if information always and only refers to other information, can then information originate in any genuine sense? Is there a perpetual-motion machine that is able to generate meaningful information in a context-free manner, that is, out of nothing? Or is the generation of information in Nature and in society actually a gigantic, context-bound process of change, recombination and re-assessment of information—one that invokes the elusive impression of the emergence of new information from nothing? And what does "new" mean in this connection? Such questions push us hard up against the edge of fundamental research, where one can see question upon question of this kind mounting up, questions about which there is room for wonderful speculation, but for which we as yet have no answers that might point the way ahead. Nevertheless, this deeper insight into the character of information has already paved the way for new concepts concerning the origin of semantic information in prebiotic matter which could enrich our present ideas on the origin and early evolution of life [20].

5.4   The Language of Genes

According to everything that we have found out about the nature of language, it seems reasonable to regard language as a phenomenon of Nature. The biophysicist Manfred Eigen expressed no doubt that "language as a principle of organisation and as a medium of exchanging information [is] a natural phenomenon" that "can be analysed in an abstract manner, without reference to human existence" [8, p. 181]. Here Eigen is referring the reader, quite rightly, to the astounding analogies between human language and the language of genes that we have already addressed (Table 5.1).

Let us examine some facts of molecular genetics. As described in Sect. 3.1, the carriers of hereditary information, the nucleic acids, are built up from four classes of nucleotide, which are placed in order in a sequence in the nucleic acid molecule, like the written symbols in a language. Just like human language, genetic information is organised hierarchically. Three nucleotides build a code-word, comparable to a word in human language. The code-words are joined up into functional units, the genes. These correspond to a sentence in human language. In turn, the genes are connected in higher-order functional units, the chromosomes, each comparable to a longish text. Moreover, the storage of genetic information comprises punctuation marks, which denote the beginning and end of a unit to be read. Last, but not least, the chemical structure of the nucleic acids also defines the direction in which the genetic information is to be read.

Alongside the parallels collated in Table 5.1, there are still further correspondences between the structural principles of the genetic and human languages. All in all, these striking parallels make it seem justifiable to speak of a "language of genes". The parallels include in particular the aperiodicity of genetic symbol sequences and the context-dependence of genetic information. The latter is provided by the physico-chemical conditions in the environment, which confer upon the—in itself ambiguous—genetic information an unambiguous meaning.

In recent years, on account of the far-reaching parallels between the structures of human language and genetic language, even the linguistic theories of Chomsky have moved into the focus of molecular biology [24]. This is associated with the hope that the methods and formalisms developed by Chomsky will also prove suitable to elucidate the structures of the molecular-genetic language. To what extent this idea will bear fruit, we shall have to see. In the most general sense it does seems likely that the application of linguistic methods to biology could open up completely new theoretical perspectives for the understanding of living matter.

Table 5.1 Hierarchical organisation of the molecular language of genes (from [10])

Unit of genetic script	Size of a single unit	Number[a] of different units in an individual organism	Demarcation[b] (punctuation) carried out by	Function	Analogous unit in language
Nucleotide	One symbol	4	Molecular structure	Primary coding symbol	Machine-symbol or letter
Codon	Nucleotide-triplet	64	Nucleotides	Translational unit (symbol for amino acid)	Phoneme or morpheme
Cistron (gene)	130–1000 codons	many thousands	Codons (Initiation, AUG; Termination, UAA, UAG, UGA)	Coding unit for protein	Word or (simple) sentence
Scripton (operon)	Up to 15 cistrons (plus intervening regions)	many thousands	Promotor (operator) Terminator	Transcription unit (mRNA)	Complex sentence
Replicon	Up to several hundred scriptons (plus intervening regions)	several (in some cases only one)	Replicator Terminator, cohesive ends (for viruses)	Reproduction unit	Paragraph
Segregon (chromosome)	Several replicons	a few (in some cases only one)	Centromere Telomere	Meiotic unit	Paragraph
Genome	A few segregons	one	Cell's nuclear membrane	Mitotic unit	Complete text
Genotype	Genome plus cytoplasmic information-carriers	one	Cell membrane	Total information	Complete text and commentaries

[a] The numbers in the first two rows are the same for all organisms, but the other numbers depend upon the complexity of the organism in question
[b] Up to and including the level of the segregon, the information is arranged sequentially. However, this does not mean that it must all be read out in a single sweep

The idea that genetic material might be structured according to the principles of a language was already posited at the end of the 19th century by Friedrich Miescher, the discoverer of the nucleic acids. He came up with this idea during his study of stereochemistry. Miescher had been fascinated by the fact that chemical compounds containing an asymmetric carbon atom have two possible molecular structures, which are related to one another in the same way as a three-dimensional object is related to its mirror-image. "In the case of the enormous molecules of the proteins", Miescher wrote in 1892 to his colleague Wilhelm His, "[…] the many asymmetric carbon atoms offer such a colossal number of possible stereoisomers that all the richness and all the variety of heredity could just as well find its expression in them as do the words and concepts of all languages in the 24–30 letters of the alphabet" [22].

Apparently, Miescher had seen intuitively that the enormous complexity of living matter presupposes a generative principle of the kind that language offers, that is, the formation of an unimaginably large variety of possible hereditary dispositions out of a few basic elements. In the trenchant words of the philosopher Hans Blumenberg, Miescher's acute observations imply that living organisms behave not like "clockwork" (*Uhrwerk*), but like "speechwork" (*Sprachwerk*) [2, p. 395 f.]. However, one should add that this "speechwork" is, as explained above, not just a rigid set of rules, but rather a creative dynamic, and as such it is a fundamental characteristic of organic matter.

Naturally, the comparison between the language of genes and that of humans breaks down if we wish to take into account the full richness of human language and use this as a measure for the genetic one. It would be more correct to view language as a phenomenon that goes through all the realms of Nature, finding its simplest expression in the language of genes and its highest form of expression (as far as we are yet aware) in human language.

The principles of the storage, transfer and generation of information lie at the root of all stages of living matter, from the biological molecule to the human being. These principles include the formation of defined rules for assignment, the aperiodicity of the coding symbols, the recombination of autonomous units of information, the context-related assignment of meaning and much more. These all contribute the characteristic features of a language without which the origin and evolution of life would presumably have been impossible. Seen from this perspective, language must indeed be considered as a general phenomenon of Nature.

References

1. Bar-Hillel, Y., Carnap, R.: Semantic Information. Br. J. Philos. Sci. **4**, 147–157 (1953)
2. Blumenberg, H.: Die Lesbarkeit der Welt. Suhrkamp, Frankfurt am Main (2000)
3. Bradbury, J. W., Vehrencamp, S. L.: Principles of Animal Communication. Sinauer Associates, Sunderland (2011)

4. Cassirer, E.: Substance and Function & Einstein's Theory of Relativity. Dover Publications, New York (1953). [Original: Substanzbegriff und Funktionsbegriff, 1910]
5. Chomsky, N.: Syntactic Structures. Mouton, The Hague (1957)
6. De Saussure, F.: Course in General Linguistics. Duckworth, London (1983). [Original: Cours de Linguistique Générale, 1916]
7. Deleuze, G.: How do we recognize structuralism? In: Desert Islands and Other Texts. Semiotexte, New York (2004). [Original: A quoi reconnaît-on le structuralisme?, 1972]
8. Eigen, M.: Self-oganization of matter and the evolution of biological macromolecules. The Science of Nature (Naturwissenschaften) 58, 465–523 (1971)
9. Eigen, M., Winkler, R.: Laws of the Game. Princeton University Press, Princeton (1993). [Original: Das Spiel, 1975]
10. Eigen, M.: Sprache und Lernen auf molekularer Ebene. In: Peisl, A., Mohler, A. (eds.): Der Mensch und seine Sprache, pp. 181–218. Propyläen Verlag, Frankfurt am Main (1983)
11. Gadamer, H.-G.: Truth and Method. Continuum, London/New York ^{2}(2004). [Original: Wahrheit und Methode, 1960]
12. Goethe, J. W. von. : Werke. Hamburger Ausgabe in 14 Bänden, Bd.13. Beck, München (1981)
13. Goethe, J. W. von.: Werke. Hamburger Ausgabe in 14 Bänden, Bd. 8. Beck, München (1981)
14. Hjelmslev, L.: Prolegomena to a Theory of Language. University of Wisconsin Press, Madison (1961). [Original: Omkring Sprogteoriens Grundlaeggelse, 1943]
15. Kirsch, R. et al.: Host plant shifts affect a major defense enzyme in Chrysomela lapponica. Proc. Natl. Acad. Sci. USA 108, 4897–4901 (2011)
16. Küppers, B.-O.: Information and the Origin of Life. MIT Press, Cambridge/Mass. (1990). [Original: Der Ursprung biologischer Information, 1986]
17. Küppers, B.-O.: Der semantische Aspekt von Information und seine evolutionsbiologische Bedeutung. Nova Acta Leopold. 294, 195–219 (1996)
18. Küppers, B.-O.: Nur Wissen kann Wissen beherrschen. Fackelträger, Köln (2008)
19. Küppers, B.-O.: Elements of a semantic code. In: Küppers, B.-O. et al. (eds.): Evolution of Semantic Systems, pp. 67–85. Springer, Heidelberg (2013)
20. Küppers, B.-O.: The nucleation of semantic information in prebiotic matter. In: Domingo, E., Schuster, P. (eds.): Quasispecies: From Theory to Experimental Systems, pp. 23–42. Springer International Publishing, Switzerland (2016)
21. McKay, D.: Information, Mechanism and Meaning. MIT Press, Cambridge/Mass. (1969)
22. Miescher, F.: Letter to W. His of 17 December 1892. In: Histochemische und physiologische Arbeiten, Bd. 1. Vogel, Leipzig (1897)
23. Patzig, G.: Sprache und Logik. Vandenhoeck & Ruprecht, Göttingen (1981)
24. Searls, D.B.: The language of the genes. Nature 420, 211–217 (2002)

Chapter 6
Can the Beauty of Nature Be Objectified?

Beauty is true, truth is beauty. This famous dictum, propounded by the poet John Keats, expresses a view that is shared by many scientists. Is there really an objectifiable connection between beauty and truth? Can one put the idea of the beauty of Nature onto a scientific foundation? Can we even find laws underlying natural beauty?

© Springer International Publishing AG 2018
B.-O. Küppers, *The Computability of the World*, The Frontiers Collection,
https://doi.org/10.1007/978-3-319-67369-1_6

6.1   Science and the Ideal of Beauty

For most people who are far from science, the activity of a scientist may seem prosaic and tedious. Only a few would associate it with the feeling of aesthetic pleasure. However, aesthetics in fact play an important part at least in the mathematical sciences. Albert Einstein is well-known to have been fascinated by physics, because he saw in physics the key to the law-like harmony of the universe. In his book "The Evolution of Physics", which he published together with Leopold Infeld, Einstein wrote: "Without the belief in the inner harmony of our world, there could be no science. This belief is and always will remain the fundamental motive for all scientific creation" [1, p. 296]. The physicist Werner Heisenberg [3] went so far as to compare the realisation of new scientific ideas to a kind of "artistic perception".

At the top of the scale, mathematicians seem to be completely devoted to the ideal of beauty. Godfrey Hardy claimed of mathematics, in all seriousness, that "Beauty is the first test; there is no permanent place in the world for ugly mathematics" [2, p. 85]. In the same sense, the mathematician and physicist Henri Poincaré [7] declared that in his work he had always allowed himself to be led by an instinct for mathematical beauty, which he described as a "truly aesthetic feeling" for the harmony of numbers, forms and geometrical elegance, one that would be known to all mathematicians. "The scientist", he elaborates,"does not study nature because it is useful; he studies it because he delights in it, and he delights in it because it is beautiful. If nature were not beautiful, it could not be worth of knowing". And he adds: "I mean that profounder beauty that comes from the harmonious order of the parts and which a pure intelligence can grasp". It is "this especial beautiy, the sense of the hamony of the cosmos, which makes us choose the facts most fitting to contribute to this harmony, just as the artist chooses from among the features of his model those which perfect the picture and give it character and life" [7, p. 366 f.].

The idea of beauty in science, as expressed here, has a long tradition. It extends back to the origins of scientific thinking in ancient times. Even in those days, philosophers developed models with which they attempted to explain the structure of the world by appeal to the ideal of beauty and harmony. The number theory of the Pythagoreans was influenced by this ideal, just as were Plato's conception of regular bodies as the basic elements for the explanation of all existence. With the beginning of modern science, Johannes Kepler took up this tradition in his attempts to explain the "harmonic" arrangement of the planetary system by reference to the Platonic bodies.

In fact, the ideal of beauty in science draws its strongest influence from the philosophy of Plato. For Plato, the idea of beauty, together with the idea of truth and the idea of the good, occupied the most prominent place in his doctrine of ideas. He regarded the unity of these ideas as the highest rung on the ladder of spiritual and divine being. In neo-Platonism, this thought finally took on mystical properties, because it was believed that the unity of ideas reflected the resplendence of the "one Good", which was shining through all phenomena and, thus, was reflected in every part of the universe. If one follows this line of thinking, the beauty of the world must be considered as an absolute truth, because the world is an image of the perfection and magnificence of its creator; the world *is* therefore *as* it is and could not be otherwise. Consequently, truth was regarded as beautiful and beauty as true.

The fact that this idea is still at home in modern science is documented by many avowals of leading scientists. In the mathematically oriented sciences, the equation "beauty = truth" seems even to be a guiding thought in the search for knowledge. The connection between aesthetic perception and scientific cognition is obviously so strong that, as Heisenberg decribed it, creative activity in science comes close to that in the arts. Interestingly, the intimate relationship between the sciences and the arts has led not only to an "astheticisation" of the sciences, but also to a "scientification" of the arts. For this, too, there are numerous examples.

The aesthetics of scientific cognition falls within the scope of "subject" aesthetics, which is concerned, in the broadest sense, with the question of how we arrive at aesthetic judgements and what part they play in the process of creative actions. In contrast, the search for objective criteria of natural beauty transcends the sphere of the subject who recognises it. Natural beauty now becomes an object of aesthetic judgements that raise, among other issues, the question of its characteristics.

It would be beyond the scope of this book to discuss whether this splitting-up into "subject" and "object" aesthetics is really meaningful, or whether the consideration of natural beauty can really be divided up at all into the subject who recognises and the object that is recognised. This would ultimately lead us into unfathomable depths of philosophical discourse.

Instead, let us adopt a phenomenological standpoint and attempt to shed some light on the connection made in science between beauty and truth. To do this we consider "natural beauty" as we believe we recognise it. In doing so, we shall not restrict our considerations to the phenomena of Nature that are accessible to our direct perception. Rather, we shall look at some phenomena of which modern science has only recently made us aware.

Using the aids of technology we can today "look" deeply into Nature and in that way indirectly "perceive" her micro- and macrophenomena that lie outside our tangible reality. More than that: by performing computer simulations we can also visualise Nature's mathematical structures. Computer-aided visualisation of that kind has, in the course of research into complex natural phenomena, given rise to images that have a high degree of aesthetic appeal and which once more raise the question of whether, and if so to what extent, "natural beauty" can be objectified and formalised.

As our ideas of beauty are strongly influenced by our feelings, and are thus to a large extent subjective, we should approach this question with the greatest caution. However, on the other hand, scientists very often refer to allegedly objective criteria such as simplicity, elegance, order, harmony and so forth, when their search for knowledge is led by the idea of beauty. Of course, the application of such criteria makes room for subjective interpretations. Nonetheless, we shall try, with all due caution, to express these criteria precisely.

To lead us along the way we make use of the working hypothesis that it is precisely our sense of simplicity, order and harmony which puts us in a position to perceive the law-like aspects of complexity and thereby to reduce the complexity itself. Thus, the aesthetic sensation could be a heuristic element in the scientific search for truth, a search that is inherently directed towards finding the simple, regular and logical order of the world. Against this background, we follow for a while the connection, emphazised above, between "natural beauty" and law-like regularity in Nature. The first step along this path will consist in classifying Nature according to its complex phenomena.

6.2 Features of Natural Complexity

The exact sciences are traditionally based on the method of simplification, which includes idealisation and abstraction. This can be seen with especial clarity in physics, which in the past concerned itself exclusively with simple systems of that kind. Only in recent decades have the exact sciences been able to develop methods and concepts that, aided by modern computer technologies, also allow the investigation of complex phenomena in Nature and society.

How complex our world actually is, we shall now demonstrate by looking at some examples from Nature (Fig. 6.1). There can be little question that the night sky shown in Fig. 6.1a is a phenomenon that reveals a staggering degree of complexity. One can well describe this kind of complexity, characterised by the random-looking distribution of stars, as "unordered" complexity. Another typical example of this kind of irregular complexity is the bizarre cloud formation shown in Fig. 6.1b.

However, many complex structures of Nature still reveal signs of order. In Fig. 6.1c a typical tree structure is shown, with its characteristic pattern of branches. Each (sufficiently large) section of this structure, in turn, contains all the principal structural features of the whole. Its overall appearance might be described as "self-similar" complexity.

Self-similar structures occur very frequently in Nature. In fact, the structure in Fig. 6.1c is not really a tree: it is the estuary region of the Colorado River flowing into the Gulf of California. Yet in the aerial photograph it looks deceptively like a tree. A further illustrative example of self-similarity is the path of an electric discharge, with its dendrite-like branching pattern (Fig. 6.1d).

Let us finally look at two examples of highly "ordered" complexity. Figure 6.1e shows, at high magnification, the compound eye of a hoverfly. This structure, too, is astonishingly complex. The facets of the eye could hardly be packed more tightly. Most impressive of all, however, is the high degree of order: the facets appear to be arranged as regularly as the atoms in a crystal lattice. No less impressive is the ordered complexity of the pollen-bearing head of a sunflower (Fig. 6.1f).

If we examine these illustrations more closely, it becomes clear that the transitions between the various forms of structural complexity are actually gradual ones. Take for example the night sky and the cloud formation. In both cases, a

Fig. 6.1 Forms of natural complexity. The night sky (**a**) and cloud formations (**b**) both show an apparently disordered complexity. Dendritic structures (**c**) and the path of forked lightning (**d**) are examples of self-similar complexity. The compound eye of a fly (**e**) and the head of a sunflower (**f**) are typical of ordered complexity

suitably selected excerpt and a suitable magnification bring certain orderly patterns into focus. The night sky reveals many ordered structures, such as galaxies with their characteristic spiral form. The cloud formation in turn shows self-similar patterns resembling those we know from mountain landscapes, trees, coastlines and the like.

Even when we take into account the fact that our perception is greatly influenced by the restrictions of perspective, the collection of pictures shown in Fig. 6.1 still conveys the impression that our sense of the beauty of these natural phenomena increases as the degree of order rises.

Before pursuing this thought further, let us take a look at one more form in which natural complexity appears—one that differs fundamentally from structural complexity. This is the phenomenon of "dynamic" complexity, characterised by its processive nature.

An instructive example of this is the dynamic complexity of the so-called Brownian motion of molecules (Fig. 6.2a). This was discovered in the 19th century by the Scottish botanist Robert Brown and was explained at the beginning of the 20th by Albert Einstein on the basis of the thermal fluctuation of particles (see Sect. 7.2). This movement of particles is a purely random affair: nothing is directed or co-ordinated. If we draw the path of a particle over a long period, it turns out to be completely disordered.

In the living cell, too, the mobile components appear at first sight to be moving completely at random. If we observe a cell under a light microscope, we have great difficulty in recognising any indications of a harmonious interplay between the various parts of the cell. This is above all because a cell is a relatively poorly co-ordinated assemblage of structures. Here, beauty in the form of harmony and order finds its expression in the "functional" organisation of the cell structures. However, in contrast to structural order, functional order cannot perceived direct. At best, one gets a hint of this when one looks at the flow diagram of the bio-chemical processes of cellular metabolism, shown in Fig. 6.2b. Even if this network of metabolic paths still looks complicated and confusing, there is nothing random or uncontrolled about the formation and decay of the various metabolic products—on the contrary, these are highly co-ordinated. All the metabolic cycles of the living cell, and the dynamics of their production, are minutely aligned. However, even the complex flow diagram gives only a superficial view of the perfect harmony of the entire interactive network of metabolic cycles.

These examples show once more that beauty and harmony cannot so easily be made manifest in the phenomena of Nature. Apparently, here too the transitions are gradual—unless one wishes to lump all Nature and all her phenomena together under the heading "beautiful" or "sublime", as in fact has been traditional in our culture for so many generations.

To characterise more exactly the appearances of natural complexity, it will be helpful to use an abstract representation of the various forms of complexity. Let us assume that natural structures are put together from individual elements according to a modular design principle. For simplicity, let us denote these elements with letters (A, B, C...). The letters will then stand for various kinds of atoms, mole-cules, cells, or for any other structural element.

An ordered complexity containing only two classes of such elements might then be represented by a simple sequence like this

ABABABABABABABABABABABABABABABABABABABAB

This representation could easily be extended to two or three dimensions. Figure 6.3 shows such a highly ordered three-dimensional structure, as encountered for example in crystals of common salt, with its strictly alternating sequence of sodium ions (A) and chloride ions (B).

(a)

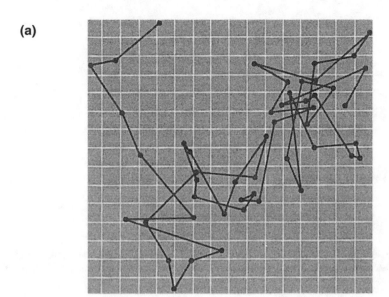

(b)

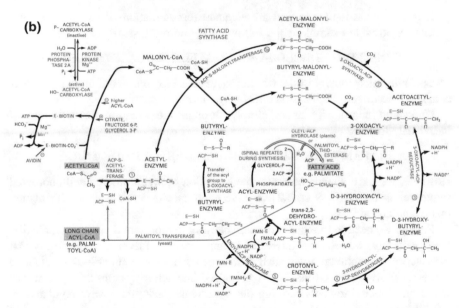

Fig. 6.2 Forms of dynamic complexity. **a** Brownian motion. Microscopically small particles suspended in a drop of water move in a highly irregular fashion. In the picture shown here, the (actually three-dimensional) movement of such a particle is projected onto the horizontal field of view of a light microscope. The lines connect the positions of the particle as recorded at 30-second intervals (from [9]). **b** Functional complexity, as expressed in the ordered dynamics of metabolic processes in cells. The excerpt shows the co-ordinated individual steps in the synthesis of fatty acids

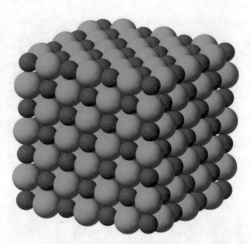

Fig. 6.3 The ion-lattice of a crystal of common salt. Sodium ions (*violet*) and chloride ions (*green*) together form a regular lattice in which each ion is surrounded octahedrally by six ions of opposite charge

It is somewhat harder to find an adequate representation of self-similar complexity. A possible example is the following palindromic sequence of letters:

EDECEDEAEDECEDEAEDEEDEAEDECEDEAEDECEDE

The palindromic property of the whole sequence is also preserved in its parts, as is characteristic of self-similar structures. (Fig. 6.4)

Unordered complexity arises simply through the random concatenation of individual structure elements:

ABBAPAMWNHBBZUDDJEIKJFAAMBBPOEJNJDIOLSJUA

However, we have already seen that the transition between ordered and unordered sequences is fluent. In Sect. 3.2 we have even specified this aspect with mathematical precision.

Finally, we need to look for some suitable way to represent dynamic complexity. Here, naturally, we are most interested in appearances of dynamic complexity that indicate some functional order, such as the ones that we encounter in the metabolic processes of living cells. Such highly organised interactive systems the dynamics of which are directed towards carrying out a particular task, are always regulated by information.

For living systems this information can be represented by a sequence of "genetic" letters, the so-called nucleotide sequence. An example is the excerpt of the genetic blueprint of the virus MS2 shown on page 39:

...UGCACGUUCUCCAACGGUGCUCCUAUGGGG...

Fig. 6.4 Virtual worlds. Mountain landscapes, coastlines, trees, river systems and the like have self-similar forms that—as the mountain range shown here—can be generated realistically in the computer, with the help of suitable algorithms (from [6]). Objects of this kind do not possess an integral number of dimensions; rather, their dimensionality is fractional. They are therefore termed "fractals"

The information appears here as a sequence of "genetic" letters (nucleotides) that is unordered in respect of its syntax. At the same time the sequence represents a meaningful piece of genetic information, as it encodes a biological structure that becomes functional in the milieu of its host cell (see Sect. 3.1).

6.3   Algorithms of Natural Beauty

We have seen that we encounter two kinds of order in natural systems: structural and functional order. The first is of spatial or geometric nature, as we find in a crystal. The second is an order in time, such as is characteristic of the functional complexity of living matter. These two kinds of order are not necessarily compatible one with the other. In particular, a high and rigid degree of spatial order and a high degree of dynamic orderedness are mutually exclusive.

The distinction between these two kinds of order provides us with a first hint of how we might objectify the properties of beauty and harmony in Nature. In this connection it will first be helpful to take a look at the linguistic origin of the word "harmony". This word has descended through Latin from the Ancient Greek word "harmonia" which means something like "proportion" or "accordance". Following this line a "harmonious" relationship in Nature could be characterised by the degree of lawlikeness with which structures and/or processes fuse together into a coherent, well-proportioned whole. Harmony is clearly the sole criterion for "natural beauty" of systems in which everything is in a state of flux.

This interpretation of harmony is a very technical one. It refers to a dynamic order in Nature which—in the words of Poincaré—"a pure intelligence can grasp". It is that kind of harmony which motivated Kepler to search for the principles of the celestial spheres. However, in the biosphere we encounter a fundamentally different understanding of harmony. The harmony of the whole appears here in its ecological balance. However, this understanding of harmony hides the fact that the biosphere is also a giant predator-prey system in which living organisms mutually cancel each other out and which owes its existence to a permanent rat race. This fact is hard to grasp if we let our emotions run free, as it shows the dark side of Nature which cannot be integrated into the concepts of beauty and harmony without debasing their meaning. Here, it becomes visible how strongly one has in science to abstract from reality if one endavours to objectify the concept of "natural beauty". Having this in mind we will continue the path we have chosen: to grasp the harmonius beauty of Nature by "pure intelligence".

Although beauty and harmony are intimately related to one another, they differ in a small, but important aspect: Beauty we primarily encounter when we look at a structure, and harmony when we grasp intuitively the workings of a co-ordinated system. Beauty and harmony are apparently indicators of the order of Nature, and they can be manifested either as structural or as dynamic order. If we follow up on this understanding of beauty and harmony, then the concept of harmony appears to be the more far-reaching of the two, as it is related to both kinds of order.

We will now come back to the connection between order and complexity. Let us first consider structural complexity. Our pre-scientific understanding of complexity already tells us that the transition between the various forms of structural complexity is a gradual one and that the complexity of a structure apparently decreases to the same extent as its order increases. However, we must be careful here, as the example of the complexity of a cloud formation shows: cloud formations, which we have described as being totally orderless, are in reality by no means orderless, even though at first sight they may appear so. On the contrary: a more exact analysis shows that a cloud formation has all the properties of a self-similar structure.

This becomes clear if we use a computer programme to create a fractal mountain (Fig. 6.15) landscape and then "photograph" the profile of the landscape from above, using various degrees of brightness to indicate its height profile (Fig. 6.5). The result is an artificially produced, fractal cloud formation, and it does not differ in any visible way from a natural structure. In an analogous manner, transitions between self-similar complexity and ordered complexity are gradual; after all, self-similarity is also associated with order.

In fact, we can even provide a quantitative measure of the connection between order and complexity (see also Chap. 3). To do this we start with the sequence that we used to represent ordered complexity:

ABABABABABABABABABABABABABABABABABABABABABAB

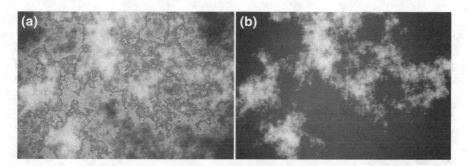

Fig. 6.5 A mountain landscape can be transformed very simply into a cloud formation (from [6]). To do this, one first generates a mountain landscape by using a suitable algorithm (similar to that used in Fig. 6.4). One then views the landscape from above, representing the different heights of the mountains by different colours, as shown on the left. The contour map thus obtained can now be converted into a deceptively realistic-looking "cloud formation" by only admitting the colours blue and white (shown on the *right*). Computer simulations of this kind show how apparently disordered structures, such as the cloud formation in Fig. 6.1b, can possess concealed regularity

As here the elements A and B appear in strict alternation, the sequence pattern can be produced by following a simple algorithm, or rule. For an *n*-membered sequence the algorithm runs:

$$(n/2) \times AB$$

This algorithm is simple, in the sense that it is more compact than the sequence itself. This is especially significant for very long sequences, in which *n* is very large. Complex sequences, by contrast, are those for which a more compact algorithm does not exist. This is apparently the case for the following sequence:

ABAAABABAAABABBBABBAABBBABAAABABAAABABBB

In this case there does not appear to be any rule with the help of which the sequence could be generated, or could be extended accurately. However, one still cannot state with certainty that the sequence is not ordered. It might well be that there is some rule (as yet unknown to us) for generating the sequence. This is a general caveat and applies to all "non-ordered" sequences. It is indeed true that there are genuinely non-ordered sequences, ones whose complexity is at a maximum; however, that can never be proved mathematically for any particular sequence. Only the reverse can be done: If, for a given sequence, a compact algorithm *can* be stated, then this proves that the sequence in question is not a non-ordered one, but rather that it is a structure with a certain degree of order.

Thus order and complexity are, in the present context, two opposite properties of a structure: the greater the order of the structure, the less its complexity and vice versa. Moreover, there are no clear borders between "ordered" and "complex"; rather, the two have to be defined according to the scientific question under consideration.

The idea of algorithmic compressibility takes account of the fact that the concepts of order and complexity possess both an objective and a subjective aspect. With the discovery of a compact algorithm, by definition, the complexity of a phenomenon is reduced, since such a compact algorithm represents a principle of order that allows a simpler description of the phenomenon. What initially appears complex—in the sense of inscrutable—is ultimately revealed to be ordered. Expressed differently: Things that appear subjectively to be complicated need not be complex in an objective sense. However, the discovery of compact algorithms presupposes much intuition and creativity, in which our aesthetic sense may also play an important part.

The fact that natural beauty implies the existence of unknown algorithms appears nowhere to be as clearly confirmed as in connection with the so-called Fibonacci series and the proportions of the Golden Section, both of which appear very frequently in Nature. The Fibonacci numbers have been known since classical antiquity, but they are named after the 13th century arithmetician Leonardo da Pisa, alias Fibonacci, who was a employed at the court of Frederic II in Palermo.

Among other problems Fibonacci was interested in the rate at which a pair of rabbits multiply. By abstracting from the natural process he derived the following rule of reproduction:

$$a_n = a_{n-1} + a_{n-2} \text{ for } n \geq 2$$

For the initial values $a_0 = 0$ und $a_1 = 1$, for example, this gives rise to the so-called Fibonacci series:

$$0, 1, 1, 2, 3, 5, 8, 13, 21, 34, 55, 89\ldots$$

This series describes the progressing size of the rabbit population. The sequence of discrete numbers in a Fibonacci series conceals an exponential law. However, the basis of this law is not the number e (2.718), but the ratio of the Golden Section (1.618). The Golden Section is a geometric measure of the continual subdivision of a line (Fig. 6.6). The proportions of the Golden Section appear whenever the length of the undivided line is greater than its larger portion by the same factor as the larger portion is greater than the smaller one. The Golden Section, to which Johannes Kepler attributed divine properties ("sectio divina"), had long been regarded in occidental art as an ideal of beauty and harmony. The proportions of the Golden Section are found in ancient Greek architecture, in art works of the Renaissance (Fig. 6.7) and in cubist paintings. No less fascinating is the observation, which goes back to Kepler, that the arrangement of the leaves of many plants stand in a direct relationship to the Fibonacci numbers and to the Golden Section (see [4]). An example of this is hoary plantain, shown in Fig. 6.8.

Now, hoary plantain is certainly not one of the most striking creations of Nature. Its beauty consists in the harmonious ordering of its leaf rosette. This, however, is not immediately perceived: it only emerges in formal representation. In contrast, the structural order of other plants, such as the arrangement of the florets of a sunflower, strikes one immediately (Fig. 6.9). It appears to us at once as a perfect order

(a) **(b)** **(c)**

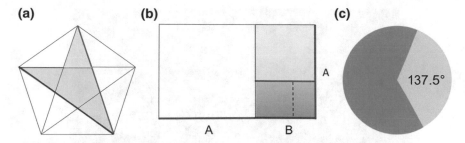

Fig. 6.6 The proportions of the Golden Section. **a** In a regular pentagon the proportions of the Golden Section are found several times. Thus, the ratio between the lengths of the *red* and the *green line* is the same as that between the *blue* and the *yellow one*, and the same as that between the *green line* and the *blue one*. Furthermore, the inscribed pentagram contains five acute and five obtuse isosceles triangles. Since the length of their legs and bases fulfil the proportions of the Golden Section, they are referred to as "Golden Triangles". **b** A rectangle with the sides $A + B$ and A has the proportions of the Golden Section when the ratio of A to B is the same as the ratio of $A + B$ to A. Such a "Golden Rectangle" can always be divided up into a square and a smaller Golden Rectangle, and the smaller rectangle can be divided up in the same way and so on ad infinitum—which reveals the self-similar character of the Golden Rectangle. **c** If a full circle is partitioned according to the Golden Section, then the larger angle that results is 222.5° and the smaller one is 137.5°. The latter is known as the "Golden Angle"

that combines beauty with harmony. Similarly, the shell of *Nautilus pompilius*, a variety of mollusc, shown in Fig. 6.10, has the proportions of the Golden Section. Moreover, the linear "struts" enclosed by the spiral form a Fibonacci series. They build the basis for perfect symmetry, which makes it possible for the organism to grow while its shape is maintained.

Wherever we find spiral arrangements—in the distribution of leaves on a twig, the arrangement of branches on a tree-trunk, the organisation of florets in a flower or seeds in a pod—we inevitably encounter the Fibonacci numbers and the proportions of the Golden Section. Kepler, and before him Leonardo da Vinci, observed the empirical connections correctly, and in our time it has proved possible, by appeal to physico-chemical models, to find an exact basis for their underlying, ordering principle (see [10]).

So far, we have concentrated upon structures that are accessible to our direct perception or to our intuition. However, there are also complex structures, with a strong aesthetic appeal, whose order remains concealed to us. These can only be conjured up with the aid of technical devices. An example is the so-called Mandelbrot set, the structures of which can be made visible on the computer screen.

The Mandelbrot set, named after the mathematician Benoît B. Mandelbrot, is a set of numbers that can be generated with the help of a simple algorithm. A number z is multiplied by itself, and the number z^2 thus obtained is increased by adding a fixed number c:

$$z \rightarrow z^2 + c$$

Fig. 6.7 The Golden Section was regarded in the ancient world and the Renaissance as the epitome of harmony, balance and proportion. There are many examples of this in architecture and the visual arts. **a** Front view of the Parthenon temple on the Acropolis in Athens. In all essential elements of its architecture we detect the proportions of the Golden Section. **b** The Apollo Belvedere. The length of his *upper* body has the same ratio to his *lower* body as the latter has to the total body length; the same is true for his other physical proportions. **c** Leonardo da Vinci's "Mona Lisa". The proportions of the picture are based upon two acute and two obtuse triangles, the sides of which show the proportions of the Golden Section

The Mandelbrot set is now generated by taking the number thus calculated as the new value of z and applying the same algorithm to it (i.e., squaring and adding c), and so on, repeating the procedure at will. For a starting value of $z = 0$ we thus obtain the series of numbers:

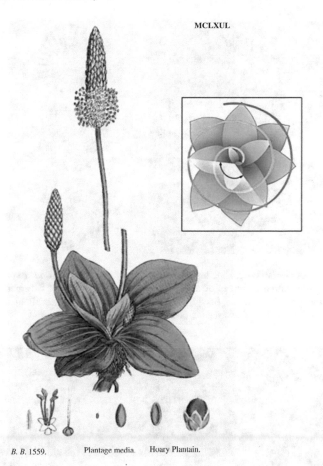

MCLXUL

B. B. 1559.   Plantage media.   Hoary Plantain.

Fig. 6.8 The hoary plantain (*Plantago media*) is one of some 250 plantain species found around the world. In its leaf rosette (inset) the leaves are arranged in a spiral, as is shown more clearly when one joins up the tips of the leaves. The angle between the medians (*mid-lines*) of two consecutive leaves (divergence angle) is always the same. Botanists state this in terms of rotations around the stalk. The numerator of the "divergence fraction" is given by the number of rotations that are needed to move spirally from one leaf to the next leaf exactly above it; the denominator is given by number of leaves that the spiral encounters on passing between these two. In our example the divergence fraction is 3/8. The most frequently observed divergence fractions found in plants are: 1/2, 1/3, 2/5, 3/8, 5/13, 8/21 and so forth. Thus the numerators and the denominators each make up a Fibonacci series

$$c, c^2 + c, (c^2 + c)^2 + c, \ldots$$

A procedure of this kind, in which the result of a calculation is used as input for the next, similar calculation, is referred to in mathematics as a *recursive* or *iterative* procedure.

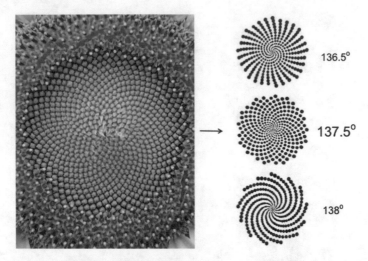

Fig. 6.9 Head of a sunflower, showing 34 *right-handed* and 55 *left-handed* spirals. These two numbers are consecutive members of a Fibonacci series. The angle of divergence corresponds fairly exactly to that of the Golden Section, as can be seen in the computer simulation of the spiral patterns for various angles of divergence (sketches *on the right*)

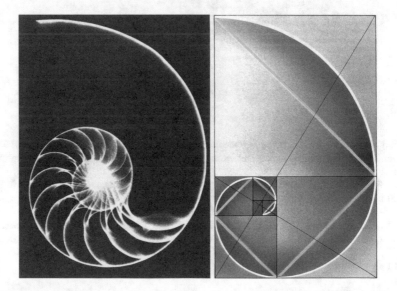

Fig. 6.10 The chambered nautilus (*Nautilus pompilius*) is a cephalopod that lives in warm ocean waters. An X-ray (*on the left*) shows its regular subdivision into chambers. Their basic geometry makes up a so-called logarithmic spiral, the exact form of which is shown on the *right*. This idealised form is obtained by recursively dividing a Golden Rectangle into a square and a smaller Golden Rectangle: if the diagonally opposite points of the consecutive squares are joined up, then the result is a Golden Spiral. The logarithmic spiral occurs frequently in Nature, presumably because it allows growth to take place while its basic form is retained: every increase in length is accompanied by a proportionate increase in radius, so that the growing structure can be enlarged without changing its shape

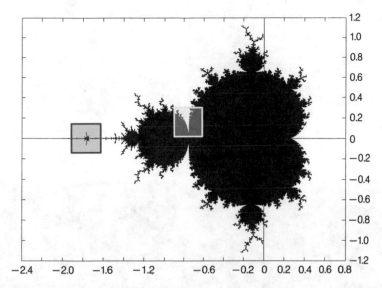

Fig. 6.11 The Mandelbrot set for $z^2 + c \rightarrow z$. The complete real part of the c plane is shown for the interval—$2.4 < \text{Re}(c) < 0.8$ and the imaginary part for the interval—$1.2 < \text{Im}(c) < 1.2$. The two coloured areas are shown in Figs. 6.12–6.17 at various magnifications. It is amazing that even such a simple algorithm can lead to extremely complex and at the same time æsthetically attractive structures (from [5])

The iteration described above leads, when real numbers are used, to relatively unspectacular results. The interesting properties are first revealed when one uses complex numbers. All complex numbers contain a real part and an imaginary part, and they can be represented as points in the so-called Gaussian plane (or Argand diagram), in which the two co-ordinates are used to plot, respectively, the real and the imaginary part of the complex number.

The set of solutions then falls into two classes. To the one class belong all the numbers that increase in every step of iteration. Such numbers, which in the course of the iteration reach a certain size, then continue to grow very rapidly and soon exceed the computer's ability to handle them. The numbers in the second class, in contrast, always remain finite, irrespective of how many steps of iteration are performed.

The Mandelbrot set in the narrow sense is the set of points that lie in the centre of the complex-number plane and includes all values of the number c that remain finite under indefinitely repeated iteration, while all the c values lying outside this set are those that are "escaping" to infinity. Points in the plane can be coloured according to their dynamic behaviour: for example, one might colour all points in the Mandelbrot set black, and all points outside it in other colours, chosen depending upon the speed with which the iteratively calculated z tends toward higher values.

The Mandelbrot set looks like a figure made up by stacking variously sized apples one upon another, which has earned it the nickname "Little Apple Man" (Fig. 6.11). However, the more interesting property of the Mandelbrot set is its edge, on which

Fig. 6.12 The edge of the Mandelbrot set (from [5]): the area indicated in *red* from Fig. 6.11 (*upper left*) is magnified (*upper right*) and an excerpt from this is shown in detail (*lower panel*)

there sit innumerable tiny forms, the shapes of which also rather resemble the Mandelbrot set. If one magnifies the edge of the Mandelbrot set, by zooming in on it with the help of a computer (Figs. 6.12–6.17), then one begins to see details of seemingly infinite variety and strong aesthetic appeal [5].

Thus, even the simple algorithm of the Mandelbrot set already leads to highly sophisticated, complex and attractive patterns, which can be made visible by computer graphics (see [6]). This findings is not only of interest in pure mathematics: it also has wide implications for our understanding of Nature, because Mandelbrot's algorithm, in various forms, lies at the bottom of many physical phenomena, such as turbulence, population dynamics or the self-organisation and evolution of matter.

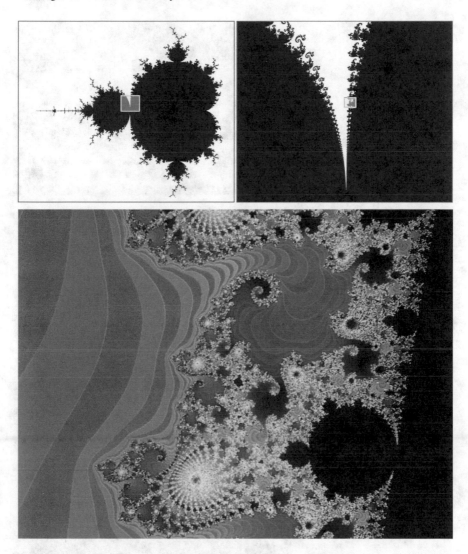

Fig. 6.13 The edge of the Mandelbrot set (from [5]): the area indicated in *yellow* from Fig. 6.11 (*upper left*) is magnified (*upper right*) and an excerpt from this is shown in detail (*lower panel*)

Alongside the discovery of Mandelbrot, numerous algorithms have been detected which are able to generate artificial structures that resemble natural ones with deceptive accuracy (see for example [8]). This shows clearly that we are on the track of the algorithmic principles of living matter, and that such matter is thus, to an increasing extent, calculable. At the same time, this opens the door to a virtual world—one in which the distinction between "natural" and "artificial" is erased and in which science and art enter into an intimate symbiosis.

Fig. 6.14 The Mandelbrot set (from [5]): an excerpt from Fig. 6.13 (*left, indicated in yellow*) is shown in detail (*right*)

Fig. 6.15 The Mandelbrot set (from [5]): an excerpt from Fig. 6.14 (*left, indicated in yellow*) is shown in detail (*right*)

Fig. 6.16 The Mandelbrot set (from [5]): an excerpt from Fig. 6.14 (*left, indicated in yellow*) is shown in detail (*right*)

Fig. 6.17 The Mandelbrot set (from [5]): an excerpt from Fig. 6.14 (*left, indicated in yellow*) is shown in detail (*right*)

6.4 What is Ultimately Beauty?

In this chapter, we have attempted to connect "Nature's beauty" with certain forms of law-like order in Nature. Nevertheless, we will obviously not succeed in ascribing objective or binding attributes to natural beauty. What is beauty, after all? We all have a sense of beauty, and can share it with one another, but even so it remains a subjective matter. This is obviously the case, as witness the enormous variety of things that people find beautiful, from the Apollo Belvedere to garden gnomes.

Yet if our ideas of beauty are subjective and their borders are blurred, then there can never be any objective criteria by which beauty can be demarcated or delimited. For that reason, beauty is determined normatively, whereby the norms that we apply depend fundamentally upon our subjective perceptions. It is indeed conceivable that our aesthetic sensibilities may one day be explained by neurobiology. However, it seems more than questionable whether our aesthetic judgements will in the same way become the objects of an exact science.

The fact that Nature is regarded as an expression of perfect beauty and harmony is surely related to the creation myth, but also to the excessive elevation of Nature in the Romantic era. However, not everything in Nature is beautiful, and not everything is harmonious. In living Nature, for example, order and regulation as an expression of harmony are the products of a long evolutionary development that in turn is based on competition, destruction and replacement.

If the phrase popular in the exact sciences, that beauty is true and truth is beautiful, is to contain even a spark of accuracy, then beauty will always remain as relative as truth. Yet, like the ideal of truth, the ideal of beauty can exert a regulative influence upon human discovery. This was the working hypothesis from which we set out. The regulative principles of the scientific search for knowledge certainly include such aesthetic aspects as simplicity, elegance and harmony. They are accompanied by the algorithmic description of Nature, through which the law-like and logically simple order of Nature is revealed. Thereby, the complexity of natural phenomena is reduced and made transparent for us. In that sense, natural beauty certainly has a guiding function for scientific knowledge—no more, but also no less.

References

1. Einstein, A., Infeld, L.: The Evolution of Physics. Cambridge University Press, Cambridge ^{2}(1978)
2. Hardy, G.H.: A Mathematican's Apology. Cambridge University Press, Cambridge (2012)
3. Heisenberg, W.: Die Bedeutung des Schönen in der exakten Naturwissenschaft. Gesammelte Werke/Collected Works. Abteilung/Series C. Piper, München (1985)
4. Huntley, H.E.: The Divine Proportion. Dover, New York (1970)
5. Peitgen, H.-O., Richter, P.H.: The Beauty of Fractals. Springer, New York (1986)
6. Peitgen, H.-O., Saupe, D. (eds.): The Science of Fractal Images. Springer, New York (1988)
7. Poincaré, H.: Science and methode. In: The Foundations of Science. The Science Press. New York/Garrison (1913). [Original: Cience et Méthode, 1908]
8. Prusinkiewicz, P., Lindenmayer, A.: The Algorithmic Beauty of Plants. Springer, New York (1990)
9. Reif, F.: Statistical Physics. McGraw-Hill, New York (1967)
10. Richter, P.H., Schranner, R.: Leaf arrangement-geometry, morphogenesis, and classification. The Science of Nature (Naturwissenschaften) **65**, 319–327 (1978)

Chapter 7
What Is Time?

A drop water falls into the ocean—a process that cannot be reversed. This irreversibility gives time a structure: past, present and future. The temporality of all happenings is surely the most mysterious phenomenon of our world. Can the exact sciences shed light on this mystery?

© Springer International Publishing AG 2018
B.-O. Küppers, *The Computability of the World*, The Frontiers Collection,
https://doi.org/10.1007/978-3-319-67369-1_7

7.1   The Thermodynamic Arrow of Time

Every process, and thus also every form of human experience, is indissolubly connected to the phenomenon of time. When we speak of time in this way, we mean first and foremost the steadily passing, measurable time, to which every material entity is exposed. Beyond that, however, the phenomenon of time has a further, more subtle aspect, one that transcends objective, physical time: it is the complex time structure that becomes manifest in our subjective sense of time. Psychological time is not a measurable quantity that passes steadily and can be measured with clocks. Rather, it is a phenomenon that arises in our conscious mind as a multilayered structure.

For the scientist, this immediately raises the question of whether the complex forms of our subjective sensation of time can perhaps be objectified and placed in some kind of relation to physical time. In addressing this question, the structure of time is of central importance: time is first and foremost a perceptible phenomenon for us because it has a structure that we experience as past, present and future. For that reason, the considerations that follow centre upon the possibilities and limitations of placing the structure of time on a scientific footing.

It will serve our purpose to start by working out some fundamental differences between "space" and "time". One of these differences, for example, consists in the dimensionality of space and time. Space is three-dimensional. In space, one can move in any of three directions, each independent of the others. Time, in contrast, is one-dimensional. The flow of time always proceeds in one direction, from the past to the future.

Yet even if we consider only a one-dimensional projection of three-dimensional space, there are still substantive differences between space and time. For example, a material particle can occupy a particular position in space at two different points in time. However, it cannot (conversely) be at two different points in space at one and the same time. Similarly, the trajectory of a point mass will always have a "past" and a "future" on the time axis, but not necessarily "left" and "right" on a space axis. It is true that we represent the structure of time in spatial categories such as that of the time axis—not least because that is the only way in which we can imagine the direction of the flow of time. Nevertheless, time is not comparable to a one-dimensional space.

Moreover, we have to admit that all these characterisations of space and time derive from the middle-sized scale of our everyday world, the so-called mesocosmos. Yet the results of Einstein's theory of relativity showed that space and time outside the mesocosmic world have completely different and unexpected properties. According to relativity theory, time does not exist independently of space; rather, together with space, it forms a four-dimensional space-time continuum. It is not impossible that similar surprises also await us in the realm of microphysics: here I am first thinking of the, as Albert Einstein termed it, "spooky action at a distance" between entangled quantum objects, which conveys the impression that a physical entity could (after all) be in two places in space at the same time.

In everyday experience no pattern of movement can be exactly "mirrored in time", that is, can be reversed by reversing the time axis. This makes it clear that the phenomenon of "temporality" has some sort of asymmetric quality. Indeed, it is the directionality of time that enables us to speak of a (temporal) sequence of states and thus to distinguish between factual (i.e., past) and possible (i.e., future) events. It is in this sense that we will use the concept of "time structure" from now on.

The temporality of all being becomes apparent in the irreversibility of natural processes. Thus, if we wish to understand the structure of time, then we must first dissect the phenomenon of irreversibility.

Irreversible processes are so familiar to us that we do not usually question them. If, for example, a drop of ink falls into a glass of water, then the ink spreads out in a cloud. At the end, the ink will become dispersed uniformly throughout the water. We observe a similar effect when a gas emerges from a small volume into a larger one: it will continue to spread until it fills the larger volume evenly. It is apparently impossible for such processes to run backwards; at any rate, no-one has ever seen a gas spontaneously contracting and squeezing itself from a large volume into a smaller one. Likewise, the water coloured by a drop of ink never loses its colour because the ink has suddenly concentrated itself back into a single drop. These processes have a key feature in common: They always lead to increased disorganisation of an initially orderly state, and never *vice versa*.

Another example of an irreversible process is shown in Fig. 7.1. Everyone will agree that the two pictures are shown in a (chrono)logical order. Nobody would entertain the idea that the process illustrated could proceed in the opposite direction. Our belief in the irreversibility of a catastrophe of that kind emerges from our experience that the generation of disorder represents the natural path of spontaneously occurring processes—not the generation of order.

Let us interpret the sequence of pictures in Fig. 7.1 in a physical way. If a motorcyclist drives into a wall at high speed, then at the moment of collision the energy of his motion (kinetic energy) is turned into other forms of energy—for example the energy causing the break-up of the wall and the motorcycle. Moreover,

Fig. 7.1 Accidents and catastrophes are painful reminders of the temporality of our world

part of the energy of motion goes into creating the very considerable disorder at the scene of the accident. Finally, a further portion of the kinetic energy goes into the molecular energy of motion of the metallic parts of the wrecked motorcycle, which are slightly warmer after the accident than before.

In view of the human catastrophe the last point sounds banal, but it is especially important for a physical understanding of irreversibility. So let us express it more precisely: The kinetic energy of a solid object (here the motorcycle) is given by the co-ordinated energy of its molecules. All these move in (more or less) the same direction, that is, in the direction of the motorcycle (Fig. 7.2 left). When the motion is stopped, for example in a collision, the co-ordinated movement of the molecules is converted to a disorganised movement, in which the individual molecules now move irregularly, to and fro along small distances, while the centre of mass of the object is now stationary (Fig. 7.2 right). What we sense as heat is precisely this irregular molecular motion.

If we wanted to reverse the transformation of ordered motion to disordered motion, we should have to impose uniformity upon the movement of the molecules. In view of the gigantic number of molecules involved, the chances of that happening on its own are as good as zero. Even the systematic unification of the molecules' movement, using some kind of experimental set-up, is only possible to a limited extent. Expressed technically: it is impossible to construct a heat-driven engine that can convert heat energy into mechanical work with an efficiency of 100%—that is, can turn the heat completely into work.

Physicists have formulated the phenomenon of the irreversibility of natural processes that lead to increased disorder as the law of increasing "entropy". This term (from the Greek *tropē*: turning or transformation) was introduced into thermodynamics by Rudolf Clausius in 1865. According to the entropy law, an isolated system—that is, a system that exchanges neither energy nor matter with its surroundings—can only undergo processes in which the system's entropy, or its disorder, increases.

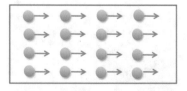 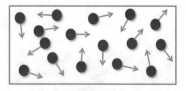

Fig. 7.2 Seen from a thermodynamic standpoint, the stopping of a body in motion turns the ordered movement of its molecules into disordered movement. In the ordered molecular movement, the energy of the system is distributed over a single degree of freedom (*left-hand* graphic). In the disordered movement, in contrast, the molecules' energy is distributed over many degrees of freedom (*right-hand* graphic). The disordered movement we sense as heat. One may regard the energy that is converted to heat as "dissipated", that is, energy which is distributed over the system in the most probable way

The law of increasing entropy is referred to as the "Second Law" of thermo-dynamics. Insofar as the universe as a whole can be regarded an isolated system, the Second Law ranks as a principle of cosmic development that characterises the direction of time. It determines the so-called "thermodynamic arrow of time". However, in the real world there are further irreversible processes, the causes of which have nothing to do with the Second Law, but which also proceed in one direction only and thus characterise an asymmetry in time. Among these are, for example, the propagation of electromagnetic waves (we only observe waves spreading out, never waves that are converging on one point), radioactive decay and the processes of evolution. However, our sense of passing time is determined exclusively by the thermodynamic time-arrow, as only this one influences the realm of everyday experience.

7.2   Weak and Strong Temporality

To arrive at a better understanding of the second law of thermodynamics and thus to a better understanding of the phenomenon of time, we shall have to express the physical concept of "disorder" considerably more precisely than we have done so far. For this purpose we need to take a closer look at the molecular structure of matter and the models available for its physical description.

An important way of characterising a thermodynamic system is by differenti-ating between its "macrostates" and its "microstates". The macrostate is a state of the system to which macroscopically measurable properties, such as pressure and temperature, can be ascribed. However, statistical mechanics show that the macroscopic properties of a system can be derived from the dynamic behaviour of its atoms and molecules. For examples, the temperature of a system corresponds to the average kinetic energy of the atoms and molecules present in it. This in turn is a result of the distribution of the speeds with which the atomic and molecular par-ticles in the system are moving.

The detailed distribution of the positions and speeds of all particles of a system at a given instant is referred to as the systems's microstate. Because of the vast numbers of particles that we encounter even in relatively small systems, we can never know the microstate of a system exactly. This can be illustrated with a numerical example: a cubic centimetre of a gas, under normal conditions, contains some 10^{19} gas molecules. So, even for this relatively small system, the description of its microstate would require 10^{19} measurements, each referring to the exact state of one of the atoms or molecules. In contrast, however, one can perfectly well make statistical statements about the system as a whole, and these statements become more reliable as the number of particles in the system increases.

Usually, a particular macrostate can be the realised by more than one microstate, i.e., more than one dynamic distribution of particles. The number of microstates that constitute a particular macrostate we refer to as the thermodynamic probability W of the macrostate. This characterisation has been chosen to express the fact that an arbitrarily chosen microstate is more likely to belong to a given macrostate when the macrostate embraces a greater number of microstates. The thermodynamic probability, for which $W > 1$ always applies, should not be confused with the mathematical probability p, for which $0 \leq p \leq 1$ invariably holds.

The number of different microstates that constitute one and the same macrostate can at the same time be understood as a measure of the order (or disorder) of a system. An ordered material system is clearly one in which knowledge of the macrostate allows one to find out, with reasonable accuracy, what its microstate is. This will obviously be the case when the macrostate only contains a single microstate. Conversely, a disordered material system is one in which knowledge of the macrostate scarcely allows any inference to be drawn about the microstate, that is, about the detailed positions and velocities of the particles that constitute it.

The second law of thermodynamics, in its qualitative version, lays down that the spontaneous development of a system always proceeds from an ordered state to a more or a less disordered one. The distinction between macro- and microstates makes this easy to understand.

Let us assume that a material system, at a particular moment, is in an ordered state—that is, in a macrostate with a low thermodynamic probability, having a small number of microstates. However, its momentary microstate is subject to perpetual change, because the particles of the system interact by colliding with one another, which in turn alters their positions and velocities. The system will thus move from its initial microstate into a new microstate, and all we know about this is that it is one of the possible neighbouring states to the original microstate. These neighbouring states will be associated with macrostates of greater or lesser thermodynamic probability. Left to itself the system will in most cases make the transition to macrostates that have a greater thermodynamic probability than that of the previous macrostate. This process will only end when the system has arrived in the macrostate with the greatest thermodynamic probability, that is, the greatest possible number of microstates. At that point the system has achieved the greatest possible degree of disordered motion, that is, of heat (cf. Fig. 7.2).

The statistical nature of the entropy principle can be illustrated by means of a simple dice game, developed some years ago by Manfred Eigen and Ruthild Winkler [5]. The model which actually goes back to a thought experiment developed by the physicists Paul und Tatjana Ehrenfest at the beginning of the 20th century makes use of a game board with $8 \times 8 = 64$ squares (Fig. 7.3). The squares are occupied by beads of various colours (in our example red and green). Each square is defined by using a co-ordinate system. By throwing two octahedral dice, we can select at random any of the 64 squares.

Before we set out the rules for the game, let us reflect briefly upon the macrostates and the microstates of the model. It clearly makes sense to regard as a macrostate the numbers of red and green beads. If we call the number of green beads n_1 and the number of red beads n_2, then the macrostate is defined by the

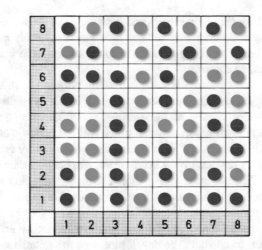

Fig. 7.3 Board game for simulating the entropy principle

number-pair (n_1, n_2). In the case at issue there are 65 possible macrostates, running from "all green" (0, 64) to "all red" (64, 0). A microstate, on the other hand, is defined by the exact distribution of beads on the board, specifying all the co-ordinates at which each colour is represented. In this case we have to list all the positions on the board and register, for each one, which colour it is occupied by: green or red. It is immediately clear that the 65 possible macrostates have different respective numbers of microstates. If n is the total number of beads ($n = n_1 + n_2$), then the thermodynamic probability W, the number of microstates contained in the macrostate (n_1, n_2), is given by the formula:

$$W = n!/(n_1! n_2!)$$

If, for example, the board is occupied entirely by beads of one colour (all green or all red), then in either case there is only one possible arrangement of the beads, and therefore only one microstate. However, the macrostate (63, 1) can be realised by 64 different possible arrangements of the beads. The greatest number of possible arrangements is achieved by an equal distribution (32, 32). In this case the macrostate comprises more than 10^{18} microstates.

And now for the game itself: In analogy to natural processes, we start from a highly ordered state and subsequently see an increase in disorder. In the extreme case, we start by occupying the board with beads of the same colour. Furthermore, we define the following rule for the dice game: In each move, the bead whose co-ordinates are selected by the dice is removed from the board and replaced by a bead of the other colour.

Played on a real board, the game is awfully boring. For this reason it is advisable to simulate it on a computer. Such simulations always end with the same result (Fig. 7.4): Independently of the starting situation, after a sufficient number of moves an equal distribution between green and red beads is obtained. However, the equipartition is a "dynamic" one: on average, just as many green as red beads leave the board, to be replaced by a bead of the respective other colour. If a deviation from the equilibrium state arises, then the probability that this deviation will be reduced immediately increases. In other words: Deviations from equilibrium are self-regulating.

The game model also demonstrates how irreversibility arises. To envisage this, one merely has to consider the development of the microstates. As the rules of the game do not favour any particular microstate, on average all microstates will occur with equal frequency. The irreversibility, which is only manifested at the level of

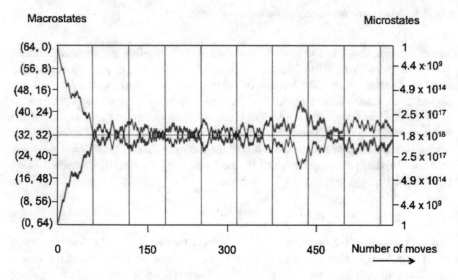

Fig. 7.4 Computer simulation of the entropy principle [5]. Starting from a highly exceptional state (only *green* beads on the board), an approximately equal distribution of *red* and *green* beads is attained within a few throws of the dice. All the microstates of the game have the same prior probability and the same waiting time for recurrence. The phenomenon of irreversibility can only be detected at the level of macrostates. It arises solely from the fact that the macrostates (as a class property of microstates) possess different numbers of microstates: the single microstate of the starting situation (64, 0) is dwarfed by the more than 10^{18} microstates of the equal partition (32, 32). As the game has a total of "only" 2×10^{19} microstates, the probability of a microstate that belongs to the even distribution is very high. The horizontal line gives a measure of the entropy of the board game. Instead of showing the discrete changes accompanying each move, the course of each coloured line is represented as a continuous path

macrostates, is due solely to the fact that the individual macrostates (Fig. 7.4 vertical axis on the left) contain different numbers of microstates (Fig. 7.4 vertical axis on the right). However, as all microstates are attained with equal prior probability, the system will in the majority of cases be found in the macrostate with the largest number of microstates. Such a macrostate will not change substantially when a distribution with the maximum statistical weight—that is, the greatest number of possibilities of realisation (microstates)—has been reached. For such a distribution to be followed by a distribution with a low statistical weight occurs only very rarely, when—as in atomic or molecular systems—the numbers involved are very large.

Ludwig Boltzmann, who was one of the pioneers of statistical mechanics, proposed that entropy be interpreted as a strictly increasing function of the statistical weight of a distribution:

Entropy of a macrostate $\sim$ Number of its microstates

With this approach, the deterministic assertion of the Second Law that the entropy of an isolated system can never decrease is put on a new footing by statistical mechanics. On the basis of the above relationship the entropy S of a system can be expressed by the following formula:

$$S = k \ln W$$

where W is the thermodynamic probability and k is a proportionality factor known as Boltzmann's constant.

The statistical concept of entropy is considerably more general than that of classical thermodynamics. This is due to the fact that in the statistical interpretation of entropy every macrostate, however far it may be from the equilibrium state, has an entropy value; in contrast, entropy in classical thermodynamics is only defined for equilibrium states.

From the statistical interpretation of entropy the conclusion can be drawn that a macrostate with a high statistical weight is only rarely followed by one of substantially lower statistical weight. For the case of a rarefied gas, Boltzmann [1] was also able to provide a function (the so-called H function) that describes the attainment of the equilibrium distribution from the mechanics of the collisions between the gas particles. As a consequence of the collision processes, the H function decreases monotonically. When its mathematical sign is reversed, the H function has the same property as the entropy function, that is, it increases in the course of time. In a state of equilibrium the H function is in fact identical (apart from a constant factor and the negative sign) to the entropy of classical thermodynamics.

Nevertheless, the statistical foundation of entropy, in particular its molecular-kinetic interpretation through the H function, has been the object of two fundamental criticisms: the "recurrence paradox" and the "reversibility paradox".

The recurrence paradox is based upon the argument that—apart from a few special cases—all the microstates of a finite material system, even the extremely improbable distributions, will return after a sufficiently long time to their starting state or to a state arbitrarily close to it. This time is termed the Poincaré recurrence time [16]. However, Boltzmann was able to show that this objection is ultimately irrelevant, because even for very small systems the recurrence time is vastly long, so that the recurrence of a microstate is in practice impossible. For example, in a cubic centimetre of a rarefied gas (at about one-thirtieth of atmospheric pressure) a microstate would only recur after $10^{10^{19}}$ years.[1]

However, it is harder to dispute the relevance of the reversibility paradox. This paradox is based on the argument that one cannot derive the irreversibility of natural events from an inherently reversible theory, such as that of classical mechanics [11]. According to the principles of mechanics, every motion can be mirrored in time, to give a possible reverse motion. Consequently, for every mechanical system in which the H function *decreases* monotonically, another system can be constructed in which all the motions of the first are mirrored, so that the H function *increases* monotonically.

Boltzmann replied to the reversibility paradox with the argument that the irreversibility of natural processes can only be derived from the principles of mechanics under certain conditions. Among these conditions is the tacit assumption that particles before collision are independent of one another, which in turn means that there is no correlation between their velocities ("hypothesis of molecular chaos"). If however, in accordance with the reversibility paradox, one constructs a system that mirrors the first in all its microphysical motions, then it would be devoid of any molecular chaos, because we would have stated from the start which particles would collide with which other ones, at what time, and with what speed.

Our discussion has now led us right to the heart of a difficult problem in fundamental physics, and this is not the place to continue along that path. Rather, let us keep in mind the result that emerges from the statistical interpretation of entropy: The macroscopic predictions made by the Second Law apply with enormously high probability, but not with absolute certainty.

According to the deterministic formulation of the Second Law, entropy is a quantity that inexorably increases with passing time. In contrast to this, the entropy game (because of the small number of "particles" involved) demonstrates the statistical character of irreversibility with especial clarity. Setting out from an extreme starting situation, the entropy function increases strongly and, once the influence of the starting conditions has diminished, fluctuates about the maximum value—that is, about the most probable value of the distribution. It is the same value which would have been expected on the basis of the deterministic theory.

A prime example of such entropic fluctuations is their appearance in the molecular movement that is known as Brownian motion. This can be directly observed in minute particles suspended in a gas or a liquid (see Fig. 6.3a). The irregular, random pattern of the particles' movement arises from the random

---

[1]This value is calculated for the reproduction of a microstate with a tolerance of 10 Ångstrøm units in space and 0.2% in the average velocity of all particles present.

collisions that they undergo with the molecules of the medium with which they are surrounded. If we focus upon a small volume of such a system then the concentration of the suspended particles fluctuates around an average value, as the chance paths traced out by Brownian motion do not always transport the same numbers of particles into and out of the element of volume that we are examining. If the average number of visible suspended particles is small, then the concentration fluctuations around the average value are relatively large and are easy to observe. The blue colour of the sky, which arises from scattering of the sun's light, likewise has its origin in fluctuations of this kind.

Our ability to see these fluctuations is more the exception than the rule. Normally, entropy fluctuations, such as those that arise when a system attains chemical equilibrium, cannot be observed. The entropy game, discussed above, also simulates that case: the attainment of equilibrium in a simple chemical reaction where molecules of substance A (green beads) are transformed with a rate constant k into molecules of substance B (red beads) and back:

$$A \underset{k_1}{\overset{k_2}{\rightleftharpoons}} B$$

In the case of equal rate constants for the forward and backward reaction (as it is realizsed in the entropy game) the law of mass action implies that a chemical equilibrium ensues in which equal concentrations of the two kinds of molecule are present (see Fig. 7.4).

Statistical theory tells us that the fluctuations around the equilibrium state have a half-width that is proportional to the square root of the total number of the particles involved. In the game model, therefore, the half-width of the fluctuations is $\sqrt{64} = 8$. Fluctuations of this size are clearly visible. However, the systems dealt with in chemistry contain vastly more particles than that. For example, a litre of a one-molar solution contains some 10^{24} molecules. The absolute size of the fluctuations is very large: $\sqrt{10^{24}} = 10^{12}$ molecules; however, the relative deviation from the equilibrium position is very small: 10^{-12}. The fluctuations thus only make a difference in the twelfth decimal place, and could not be detected with today's most sensitive instruments. That is the reason why chemists can treat the law of mass-action as a deterministic law although in fact the concentrations of molecules in a chemical reaction system at equilibrium are continuously fluctuating about their equilibrium values.

Manfred Eigen has proposed that temporality of an isolated system be termed "weak temporality", as its time-dependence disappears with the attainment of equilibrium [6, p. 213]. Accordingly, the time-dependence that arises from the reversal of an equilibrium must show "strong temporality". A reversal of this kind is not possible in isolated systems, on account of the restrictions imposed by the Second Law, but it could take place in open systems, that is, in systems which exchange energy and matter (or both) with their surroundings. Here, under certain conditions, fluctuations out of equilibrium can reinforce themselves spontaneously and—in a reversal of the entropy game—manifest themselves at the macroscopic level. In situations of this kind, the temporality of the system arises through, and during, its temporal development.

7.3   Cosmological Foundations of Temporality

We have repeatedly underlined the fact that the irreversibility of natural processes cannot be derived from the reversible laws of mechanics. Thus, irreversibility can at best be a consequence of the particular conditions under which the laws of mechanics work in Nature. Boltzmann [2] suggested two possible scenarios for this. They are based upon specific assumptions about the state of the primaeval universe, and are referred to respectively as the "initial-state hypothesis" and the "fluctuation hypothesis".

According to the initial-state hypothesis, the universe emerged from a highly improbable "initial" state. From this state until now the universe is assumed to be moving towards thermodynamic equilibrium. This one-way transition—that is the conclusion drawn from the initial hypothesis—defines the direction of time.

According to the fluctuation hypothesis, the universe in its entirety is already in a state of equilibrium, while our "local" world, in which we live, is at present in the midst of a huge, but locally restricted entropic fluctuation. The dissipation of this fluctuation leads, as the fluctuation hypothesis concludes, to a directional course of natural events, which we perceive as the asymmetry of time.

The difference between the two hypotheses is illustrated in Fig. 7.5, which shows the notional course of the entropy of the universe. The initial-state hypothesis provides a foundation for the structure of time by the assumption of exceptional starting conditions. This assumption corresponds to the starting situation in the entropy game (compare Fig. 7.4). The fluctuation hypothesis, in contrast, attempts to relate the structure of time to the portion of the entropy curve that is already in equilibrium.

Let us look at the fluctuation hypothesis in more detail (Fig. 7.6). In this case, the universe is regarded as an isolated system that is completely symmetrical in time and space. This means that there are no preferred directions in time or space. Moreover, it is assumed that the universe (considered as though it was a gas) is in thermodynamic equilibrium, that is, in a state of greatest possible disorder. However, there are fluctuations around the equilibrium state, like those we have already seen in the entropy game (Fig. 7.4). Thus, there are regions of time and space in which the degree of order deviates from the greatest possible disorder. If one considers the cosmic vastness of time and space, then it seems perfectly probable, as Boltzmann argued, that entropy fluctuations in the universe sometimes arise on the scale of entire stellar systems ("individual worlds").

Yet how are we to explain that precisely we humans inhabit such a time-structured "individual world" and that we can even perceive the passage of time? An immediate answer would be that life is in any case only possible in a world that is far from thermodynamic equilibrium. Only under those conditions can

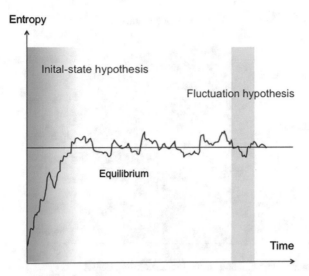

Fig. 7.5 Hypothetical course of entropy in time for the attainment of thermodynamic equilibrium in the universe. The initial-state hypothesis bases the existence of temporally structured reality upon the state of development of the universe far from thermodynamic equilibrium. The fluctuation hypothesis, in contrast, derives the structure of time from a gigantic, but local entropy fluctuation of a universe that overall is in thermal equilibrium. Both hypotheses are due to Ludwig Boltzmann

the differences between states of matter be great enough to allow life to arise. Such a situation is clearly given in the two flanks of a huge entropy fluctuation. Seen in this way, it is no great surprise that we live in just such an individual world.

One might well object that the entropy curve sketched in Fig. 7.6, in accordance with our assumptions, is symmetric. This means that the probability that we find ourselves in a phase of decreasing entropy is just as great as that of our living in a phase of increasing entropy. However, the two processes—although they run in opposite directions—define the direction of time; can one then speak at all of a past and a future? Boltzmann's reply to this question was that the definition of the direction of time is a mere consensus issue: "For the universe, the two directions of time are indistinguishable, just as in space there is no 'up' or 'down'. However, at any point on the Earth's surface, we call the direction towards the centre of the Earth 'down'; in just the same way, an organism that happens to be in a particular time-phase of an individual world will define the direction of time as being the one that is opposed to less probable states rather than more probable ones (the former would be the past or the beginning, the latter would be the future or the end) "[2, p. 396].

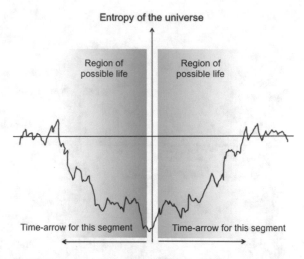

Fig. 7.6 Schematic representation of the fluctuation hypothesis

The philosopher Karl Popper criticised the fluctuation hypothesis as completely unjustifiable subjectivism. He wrote: "I think that Boltzmann's idea is staggering in its boldness and beauty. But I also think that it is quite untenable, at least for a realist. It brands unidirectional change as an illusion. This makes the catastrophe of Hiroshima an illusion. Thus it makes our world an illusion, and with it *all our attempts to find out more about our world*." [13, p. 186].

It is a matter for debate whether Boltzmann here, as Popper claims, is propagating naked subjectivism (cf. [4]). More important is the accusation levelled by some physicsts that a thoroughgoing application of probability theory to the fluctuation model leads to contradictions [3, 14].

The following consideration appears to reinforce this accusation: According to Boltzmann, the world in which we live arose through a huge entropy fluctuation. In fact, there are numerous records, such as the fossils, that indicate that our world has been passed through a long history of physical development. Thus it can no longer be in its initial state of very low entropy. Yet, according to Boltzmann's theory, an entropic state close to equilibrium—a small fluctuation—is very much more probable than a large one such as must have preceded the origin of our world. Therefore, the universe must contain many more individual worlds, whose initial state corresponds to the present, momentary state of our own world. In that case it is also more probable that all the records of our world arose by a random fluctuation than that the previous state of even lower entropy, which we believe our records indicate has existed, actually came about. Thus, according to the fluctuation hypothesis, it is far more probable that the present represents the entropy minimum and the past, which on the basis of various records we believe to have existed, is an illusion—a conclusion that would extend the fluctuation hypothesis *ad absurdum*.

Let us now take a look at the alternative explanation, die initial-state hypothesis. In our entropy game, we saw that the irreversible (overall) course of the game was a consequence of the special starting conditions. We set out from an initial state of very low entropy, so that the game inevitably developed in a direction of states with increasing probability, that is, with greater entropy. If this model is applied at the cosmic level, then this leads to the idea that the initial state of the universe must have been a highly improbable macrostate with a minimum of entropy. For example, one might imagine that at the beginning all matter was concentrated in a very small volume. If we presuppose an absolute minimum of entropy for the initial state, then this leads in the early phase to a permanent increase in entropy and thus to an unambiguous structuring of time into past, present and future.

The initial-state hypothesis thus rests upon the assumption that the universe, as the physicist Friedrich Hund expressed it, "is still very young" [7, p. 140]. Concrete models for this hypothesis include models for the expansion of the universe that have been put forward as solutions to Einstein's equations of gravity. The expansion of the universe begins here with a state of infinite density, the so-called big-bang singularity. This singularity is at once the beginning of space and time.

If we look at the history of our universe, then physical time does not, after all, seem as immutable as we had imagined. Rather, there seems to exist a "cosmic horizon" at which time "rose" and will possibly one day "set". The question of whether the universe will ever move from its present phase of expansion and enter a phase of contraction, ending ultimately in a final "big-rip" singularity, depends upon the total mass of the universe. At present this question is still unanswerable.

The initial state hypothesis is only plausible if the unique starting condition of the universe that it postulates did actually exist. Only on this basis can the universe have begun to develop in such a way as to determine a direction of time. Again, a history of the universe's development can only be inferred from records. One example of this is cosmic background radiation, which has been interpreted as the electromagnetic "echo" of the Big Bang. However, such observations are only relevant as records when we presuppose the existence of a time-structure and thus also the possibility of an objective distinction between events that have definitely took place and those that only might have. Yet the existence of the structure of time was precisely what we wanted to derive from the initial-state hypothesis.

It thus appears that we have stumbled into a circular argument: on the one hand we base the phenomenon of temporality upon our experience of irreversibility, which leaves us in no doubt that natural events possess direction, expressed physically in the Second Law; on the other hand there is little doubt that the structure of time represents an absolutely essential for any kind of human experience, not least for the experience of irreversibility that has lead in turn to the formulation of the Second Law. How can we escape from this dilemma?

The answer is: with the help of a consistency consideration. Since it is not possible to derive the existence of a structure of time from physical principles, we can only attempt to show that, under the assumption of the validity of the Second Law, the structure of time can be what we have assumed it to be.

A consistency argument is not a circular one, as illustrated by a simple example from logic. If two statements A and B are logically equivalent, then their consistency means that one can deduce both B from A and also A from B. This only turns into a "vicious circle" when the consistency is (wrongly) claimed as proving that A and B are true. All it proves is that A and B are either both true or both false.

With respect to the explanation of the structure of time, Carl Friedrich von Weizsäcker [15] attempted to derive a proof of consistency by appeal to a sharper definition of the concept of truth. He was able to show that the irreversibility of natural processes can be reconciled with the temporal symmetry of the laws of mechanics under a specific condition: that in the statistical interpretation of the entropy law probability is only used to calculate real transitions—that is, transitions toward the (respective) future. From this follows at first the increase in entropy for the future. Since, however, every moment in the past was once the present, the increase in entropy follows for everything that at that time belonged to the future— that is, also for times that today belong to the past.

The demonstration of consistency merely makes use of the assumption that the concept of probability can only be applied meaningfully to future, that is, "possible" events. However, it is meaningless to ask about the probability of "factual", that is, past events. The temporally asymmetric application of the probability concept, based as it is upon the reality of records, is here the point at which the structure of time feeds into the statistical foundations of the Second Law.

7.4 The Entanglement of Time Modes

We have seen that the structuring of time into past, present and future cannot be derived directly from the basic laws of physics. Instead, this is done at the price of additional assumptions, which however are epistemologically unsatisfactory or at least problematic. Nevertheless, everything points to temporality as a property of natural events that is present a priori. Even so, if we ascribe the irreversibility of natural processes to an objectively given time-structure, then this has an important consequence for physics: under these conditions the reversibility of its fundamental theories demands an explanation, while irreversible theory is precisely what one would expect.

It can indeed be shown that the asymmetry of time implies the reversibility of the fundamental theories. This is associated with a fundamental principle of quantum field theory that is known as the CPT theorem. According to this theorem, the causality of natural processes would be violated if it were impossible to convert a natural process by the combined application of the following three operations into another possible process:

1. Reversal of all charges ("**Charge**"),
2. Spatial mirroring about a point ("**Parity**"),
3. Reversal of all motions ("**Time**").

This in turn means that the compatibility between the elementary natural laws on the one hand and time-structured reality (and the causality principle that rests upon it) on the other comes at the price of a symmetry that swaps the past with the future [8]. In other words: The validity of the causality principle enforces, owing to the CPT theorem, the reversibility of the fundamental laws.

Finally, we take up the question of the relationship between physical time and our "inner" time, that is, psychological time. The key to this question is our experience of time. Here, the principle of causality and thus the asymmetry of time play a crucial role. Without the structuring of time into modi such as "before" and "after", causal statements—for example, "Lightning is followed by thunder"—could not be made.

For Immanuel Kant, who penetrated the presuppositions for human perception to a depth that no other philosopher has attained, time belongs to the a priori forms of human intuition through which we perceive the world and which make experience possible at all. According to Kant, "time is merely a subjective condition of our (human) intuition (which is at any time sensory, insofar as we are affected by things), and in itself, outside the subject, it is nothing. Nevertheless, in view of all phenomena, and therefore of all things, that we may encounter in our experience, time is necessarily objective" [9, p. 82]. In a footnote he once more explains: "I can indeed say that my perceptions follow one another, but that only means that we are conscious of these as taking place in some order in time, that is, according to the form of our internal senses. Time is thus not an entity in itself, and neither is it an objective characteristic associated with things" [9, p. 84].

This interpretation may suggest a subjectivistic conception of time, one which stands in the way of the attempts that we have made to find an objective justification of the structure of time. However, any such contradiction is only an apparent one. It is immediately resolved when we examine Kant's thesis in the light of evolution. The ability of the brain to recognise and to perceive, which makes us capable of acquiring experience, is certainly the result of a long evolutionary development in the course of which the functions of the brain have been ever better adapted to the world in which we live. Among the central tasks of the brain is that of taking up information about the external world and processing it, in order to ensure the survival of the organism. This task can only be carried out if the brain's internal picture indeed reflects the structure of the external world. This allows us to proceed on the assumption that time is really structured in an objective sense, and that this aspect of the external world is depicted realistically within our brain. In the light of evolution, Kant's a priori of the individual reappears as an a posteriori of our tribal history [10].

Needless to say, our awareness of time is far more complex than can be expressed on the one-dimensional scale of "past, present, future". This raises the question of whether we can ever transcend physical time and understand, in a scientific sense, the immeasurable plenitude of our subjective sense of time. Brain research seems likely to keep us waiting for this for a long time. However, we can shed some light on the problem with a few basic thoughts. If the postulate that experience is only possible within a time-structured reality is correct, then this must also apply to the experience of time itself.

However, this means that the experience of temporality is itself a time-structured process. For example, referring to the experience of the present it makes sense to speak of the "past of the present", the "present of the present" and the "future of the present" (cf. [12]). The same applies, analogously, to the experience of the other two modi of time (past and future), so that altogether nine entangled first-order modi of time arise.

However, the temporal being that manifests itself in these first-order "interleavings" is an existence that can as well be experienced within three-fold temporality. Thus 27 second-order entanglements arise, which each contains three tenses. The iterative application of the experience of temporality can be repeated arbitrarily often, and this finally leads to a multidimensional temporality, which characterizses our subjective awareness of time. The one-dimensional projection of this temporal structure corresponds to physical time in the threesome of tenses: past, present and future. Only this part of time's structure can be quantified and measured. The question of the ontological interpretation of the entanglements of the time modi opens a wide field for the philosophical investigation of the temporality of human existence.

References

1. Boltzmann, L.: Weitere Studien über das Wärmegleichgewicht unter Gasmolekülen. Wiener Berichte **66**, 275–370 (1872)
2. Boltzmann, L.: Zu Hrn. Zermelo's Abhandlung "Ueber die mechanische Erklärung irreversibler Vorgänge". Ann. Phys. **60**, 392–398 (1897)
3. Bronstein, M., Landau, L.: Über den zweiten Wärmesatz und die Zusammenhangsverhältnisse der Welt im Großen. Phys. Z. Sowjetunion **4**, 114–119 (1933)
4. Curd, M.: Popper on Boltzmann's theory of the direction of time. In: Sexl, R., Blackmore, J. (eds.): Ludwig Boltzmann Gesamtausgabe, Bd. 8, pp. 263–304. Akademische Druck und Verlagsanstalt, Graz (1982)
5. Eigen, M., Winkler, R.: Laws of the Game. Princeton (1993). [Original: Das Spiel, 1975]
6. Eigen, M.: From Strange Simplicity to Complex Familiarity. Oxford University Press, Oxford (2013)
7. Hund, F.: Grundbegriffe der Physik. Mannheim (1969)

8. Jost, R.: Erinnerungen: Erlesenes und Erlebtes. Phys. Bl. **40**, 178–181 (1984)
9. Kant, I.: Kritik der reinen Vernunft. Werkausgabe, Band III. Insel, Wiesbaden [2](1956)
10. Lorenz, K.: Kants Lehre vom Apriorischen im Lichte gegenwärtiger Biologie. Blätter für Deutsche Philosophie **15**, 94–125 (1941)
11. Loschmidt, J.: Über den Zustand des Wärmegleichgewichtes eines Systems von Körpern mit Rücksicht auf die Schwerkraft. Wiener Berichte **73**, 128–142 (1876)
12. Picht, G.: Die Zeit und die Modalitäten. In: Dürr, H.P. (ed.): Quanten und Felder, pp. 67–76. Vieweg, Braunschweig (1971)
13. Popper, K.R.: Unended Quest. Routledge, London/New York (1992)
14. von Weizsäcker, C.F.: Der Zweite Hauptsatz und der Unterschied von Vergangenheit und Zukunft. Ann. Phys. **36**, 275–283 (1939)
15. von Weizsäcker, C.F.: Unity of Nature. Farrar Straus Giroux, New York (1980). [Original: Einheit der Natur, 1971]
16. Zermelo, Z.: Ueber einen Satz der Dynamik und die mechanische Wärmetheorie. Ann. Phys. **57**, 485–496 (1896)

Chapter 8
Can History Be Condensed into Formulae?

Wars and the destruction that they bring—here the total destruction of Dresden in 1945 —are violent caesurae in human history. One goal of a general theory of history could be to model in some way the structural changes that accompany historical upheavals. But what are the determinants of historical processes? Is it conceivable that the unique course of history is determined by laws? What kind of explanations may be expected with regard to the overwhelming complexity of historical events? In short: are there any precise answers to the unending quest concerning the essence of history?

© Springer International Publishing AG 2018
B.-O. Küppers, *The Computability of the World*, The Frontiers Collection,
https://doi.org/10.1007/978-3-319-67369-1_8

8.1 History and Causal Analysis

In his collection of essays on the philosophy of science, the sociologist Max Weber speaks metaphorically of history as being a "gigantic, chaotic stream of events" that "lumbers along through time" [8, p. 214]. It is indeed true that nothing seems less ordered, and more random, than the general course of history. One can posit numerous very plausible reasons for this, of which only the three most important ones will be mentioned here.

First of all, history does not seem to obey any general laws or rules that might lead it into repeatable, and thus calculable, patterns of development. Secondly, the course of history is to a large extent determined by the free-will decisions of humans, and these decisions imprint upon history the character of uniqueness and unforeseeability. Finally, historical processes usually have so many and various aspects that it would seem impossible to isolate the causes of any particular event or to base any kind of causal explanation upon these.

It is therefore no surprise that the complexity of historical contexts makes history itself appear as an intransparent network of decision processes that, to the outside observer, seems to be "chaotic". Is there any reasonable hope that we will ever be able to calculate the course of historical processes? Will we—one day—come into possession of a set of formulae that express world history?

Clearly, we are not trying to realise an ancient dream of mankind by setting out on the search for general laws governing world events—laws of which we cannot even be certain that they exist. What follows will be aimed towards the more modest goal of finding a conceivable path towards placing the historicity of the world within a context such as to make these laws accessible within the framework of the exact sciences. We shall thus be dealing exclusively with the methodological and conceptional premises from which one might proceed towards constructing a general theory of historical processes.

Even if historical processes seem to be hopelessly complex, this need not in itself discourage us from attempting to analyse them, and providing a theoretical basis for them, on the basis of scientific methods. In systematic historical research, efforts have repeatedly been made to go beyond merely describing historical processes by taking the further step of trying to explain them within their complex contexts. However, in these attempts the term "explanation" has often been given a meaning that has only little in common with the traditional understanding of explanation in the causal-analytical sciences.

For example, historical research has often emphasised—following the position adopted by the philosopher Wilhelm Dilthey—that the scientist who researches into history is at the same time a participant in that history and takes part in forming it. This, however, would be a break with a fundamental principle of objective science, according to which the subject who acquires knowledge must maintain a critical distance from the object about which knowledge is being acquired. Some philosophers of history have believed that this problem could be bypassed by abandoning the concept of explanation as used in the causal-analytical sciences and replacing it by the concept of understanding. According to their view, the real goal of research into history is to describe and to evaluate the sense and the meaning of human actions in the past, but not to search for some kind of ostensible causal connections between individual events.

However, by moving the goalposts of knowledge in this way, historical research mutates to a kind of philosophy that is based upon a "non-analytical" understanding of history. This would mean that the course of history is now to be seen not as an objectifiable causal context, but rather as a "document" of our lifeworld, the sense and the meaning of which is actually to be dissected and expounded like a Bible text or a legal codex. With this, any understanding of history falls within the realm of hermeneutics, that is, the science of interpreting texts.

In the writings of German philosophers such as Martin Heidegger, Hans-Georg Gadamer and others the hermeneutic approach even acquired an existentialistic dimension. According to Gadamer [1], historical being must be conceived of as a coherence of human experience that has to be explained by an understanding of the "rootedness" of the human being in its existence ("Dasein"). Heidegger [3], on whose work Gadamer's rests, speaks of the "Being-in-the-World" as the most primal essence of human existence, and of understanding as the most fundamental act of its performance.

An approach of this kind, which proceeds from the unity of subject and object within the act of understanding, and according to which historicism and history are reciprocally interrelated, is very difficult to bring into assonance with the goals of analytical science. If, on the other hand, we wish to retain the causal-analytical concept of explanation, then clearly we must approach the problem of historical processes with a completely different strategy. In particular, we shall not attempt to grasp and to explain from the start the phenomenon of historicity in all its complexity. Rather, we will proceed analytically and examine whether, and if so to what extent, the complexity of the problem can be reduced by appropriate abstractions and idealisations.

Thus, to begin with, we can ask what the minimum conditions are that characterise a historical process and then proceed to study the general properties of such processes under these conditions. Only in this way can we reconstruct the original complexity of the problem. In adding step by step more aspects into the concept of the historical process, we attain an increasingly sophisticated picture of the course of history, one that finally gives a view of history that is largely true to reality.

The reductionistic approach to science that we have now arrived at has shown its value again and again, especially in the natural sciences. When, for example, in the 18th century people began building steam engines, they seemed to be confronted with an unsolvable physical problem. The basic laws of mechanics were reversible —that is, any process (it was thought) could be turned around. They did not include any preferential direction with respect to time. However, a technical process is always an irreversible one, because it represents movement towards a desired end or goal. To bring the physics of reversible processes to bear upon a technical interactive device such as a steam engine the French engineer and physicist Sadi Carnot developed the idea of an idealised reversible steam engine, the so-called "Carnot engine", in which all processes take place reversibly. In this way he succeeded in calculating important working quantities such as the efficiency of the engine, irrespective of the fact that all processes in a real steam engine take place irreversibly.

How serious Carnot's idealisation of the reality actually was is illustrated by the following consideration. The statement that a process takes place reversibly implies that the system is at equilibrium at all times. This means that the system can change its state reversibly only by passing through an infinitely dense sequence of equilibrium states. Under these premisses, reversible changes in state are only possible if they take place infinitely slowly, so that the system never gets out of balance. However, no machine in the world can operate like that. Reversible processes are therefore idealisations and, strictly speaking, they never happen in the real world.

Nevertheless, Carnot worked with idealised, imaginary models in which steam engines were treated as systems in equilibrium. By doing this, he could calculate their properties, even though he knew perfectly well that no steam engine could ever work under such conditions. This example is not unique. Most models in physics are based principally upon idealisations and abstractions that in reality never exist. These include such important concepts as the "point mass", "point charge", "ideal gas", "frictionless movement", "elastic collision" and the "isolated system"—to name but a few examples.

In the light of the numerous idealisations and abstractions which characterise the methods of the exact sciences one cannot but be surprised by the fact that these are nevertheless able to explore reality with the greatest accuracy. Why, one may very justifiably ask, should idealisation, abstraction and simplification not be sound methodological tools in studying historical processes? Because there are no general laws of world history? Because the influence of deliberate, goal- and purpose-directed actions poses an insuperable barrier for the causal-analytical understanding of historical processes? Or perhaps because the causal-analytical method is from the beginning doomed to fail on account of the complexity of the problem?

Let us rehearse the counterarguments, one at a time. First of all: Are there really no general laws governing world events? This question cannot immediately be pushed to one side as being unrealistic or even meaningless. Before doing that one must clarify what we mean by "world events". This is an extraordinarily far-reaching concept. It includes not only the history of human actions, but also the history of Nature. There indeed seem to be universal laws governing the development and thus the history of Nature. One need only think, for example, of the theories of the development of the universe or of the theories of the origin and evolution of life.

A harder, many-layered question to answer is the second one: that of deliberate actions by human beings and the influence that these have upon the course of human history. The dark side of such actions was illustrated in a terrifying way by the painter Matthias Grünewald in the 16th century (Fig. 8.1). However, here too we can ask the counter-question: How great in reality is the influence of "free-will" decisions upon history? Anyone who has been caught up in a panicking crowd, or been the victim of a vilifying rumour, or been exposed to the suggestive power of a mass event, knows how strong the influence of a crowd of people upon the opinion, the decision-making and the action of an individual can be.

Such mass crowd effects, in which the intentions of an individual are reinforced or suppressed by co-operative mechanisms, do appear to play a substantial part in the events of history. Without such amplification or control mechanisms, historically insignificant individual decisions would hardly be noticed. Yet mass crowd phenomena and co-operative interactions can be well explained by application of the causal-analytical approach.

Finally, what weight should be attached to the objection that historical processes are much too complex to be analysed by the exact sciences, because they have a virtually infinite number of causes? This is tantamount to saying that one can only

Fig. 8.1 The church altar in Issenheim (Alsace, France), where Matthias Grünewald depicted the temptations of St. Anthony. Violence and destruction, which here are shown as proceeding from monstrous creatures, belong to the incalculable dark side of human activity. Does this put history beyond a causal-analytical understanding?

distinguish between primary and secondary causes on the basis of value-judgements and, moreover, that one cannot even be sure of having captured completely all of the causal factors.

However, strictly speaking, that argument applies to each and every phenomenon in the world. Ultimately, the entire world is the cause of every event that takes place within it. However, since we cannot really practice science by repeatedly investigating the entire world, we must isolate events in some appropriate way from the rest of the world. The preparation of the object of knowledge depends fundamentally upon the scientific question being asked, the question with which we intend to approach reality and to explain the character of events within it. Here we do not need to seek out ultimate causes—it suffices to discover the causal determinants that are relevant within the little section of the world that we have identified as being relevant. A science whose tools in trade comprise abstraction, idealisation and simplification will certainly be aware of the relative nature of its discoveries and will thus avoid foundering upon the complexities of reality.

Even if it might appear impossible for historical events ever to be calculable, there are indubitable first signs on the scientific horizon that suggest that there may be some hope of this. Recent results in physics, in particular from chaos theory and the theory of self-organisation, have allowed a profound look into the law-like behaviour of unique development processes. Furthermore, the applications of game theory to complex human decision situations have shown that the effects of intentional action are also calculable to a certain extent. Finally, sciences such as biology or meteorology have given us impressive examples of how historical processes in Nature can be of stupefying complexity and nonetheless amenable to application of the causal-analytical method.

In the light of this progress it may well prove, one day, a meaningful task to investigate the extent to which the phenomenon of historicity can be modelled with scientific rigour. Put more exactly: The task is not to explain historical events, manifold and unique as they are, in a law-like manner. Rather, it is (exclusively) to illustrate the nature, the characteristic features and the structure of historical events from the perspective of the exact sciences, as we shall now attempt to do.

8.2   The Essence of Historicity

The most fundamental property of our world is its temporality. It has left deep traces in innumerable historical records. Yet, the essence of historicity is by no means exhausted in its temporality. Just as essential is the fact that historical events can never be repeated in all their details. In short: Apart from their temporality the uniqueness, individuality and particular nature of historical facts make up the second conspicuous feature of historicity. Since every event and every process in the world has these characteristics, everything that happens is ultimately historical in essence. These are elementary statements, ones that belong to the basic set of insights that together make up human experience.

As already mentioned, it is all the more surprising that physics as a science bound to experience, uses mathematical models in a way that implies the existence of processes that are in the strictest sense ahistoric, that is reversible and reproducible. The idealised assumption that natural events are reversible simply blankets out the fact that all happening in the world is inherently historical. Instead, the temporality of the world is negelected and time appears in the laws of mechanical physics as a directionless mathematical parameter that acts as a mere counting measure.

Since, in reversible physics, the temporality of the world, and with it the distinction between past and future, is completely abandoned, a tile falling from a roof can at any time reverse its direction and fly back to where it started from. At least, the equations of motion that we know from classical mechanics allow the possibility of such a surreal world. Even though these processes are in crass contradiction to what we know from everyday experience, the physics of reversible processes has nonetheless proved its use time and again in describing reality.

The concept of reversibility is associated with a picture of Nature that is symmetric in time and therefore without any history. Only with the discovery of the second law of thermodynamics in the 19th century came the first formulation of a principle that takes account of the irreversibility, and thus the temporality, of natural processes (see Chap. 7). The Second Law states that in an isolated system—that is, one that exchanges neither energy nor matter with its surroundings—only those processes can take place in which the entropy of the system increases. The entropy principle therefore defines the direction of processes that occur sponteously. Moreover, the amount by which the system's entropy increases provides a direct measure of the irreversibility of such processes. In this way the temporality of physical happenings is expressed by the thermodynamic arrow of time.

Let us come back to the general characteristics of historicity. Besides their temporality, historical events are also characterised by unrepeatability, and thus uniqueness. And it is precisely this property that has seemed to evade any kind of law-based explanations of history until today.

We must expand a little on this important point. According to our traditional way of thinking, law and regularity, or rather reproducibility, are inseparably linked together. At least this seems to be the view taken in the exact sciences, where the regular behaviour of natural phenomena is seen as the most basic property of a law-like association. But is that really the case? Let us consider physics; its central methodological instrument is the experiment conducted under controlled conditions. More than that: the physics experiment is not only the epitome of the generation of reproducible processes, but is at the same time paradigmatic for the logic of scientific explanations. The basic structure of such explanations is dualistic. On the one hand, they contain statements relating to the general laws that determine the phenomenon under investigation. On the other, they also contain statements about the particular circumstances or conditions that have preceded the event that is to be explained. For this reason such statements are referred to as "antecedence" conditions (from the Latin *antecedere*, to precede).

From a formal point of view, a scientific explanation consists in deducing the event that is to be explained from the laws involved and by reference to the antecedence conditions (for details see Sect. 9.3). If the event in question has already taken place, then we call the deduction an "explanation". If it lies in the future (such as the occurrence of a solar eclipse), we refer to it as a "prediction".

The physics experiment is precisely the practical realisation of this model of scientific explanation. Here, matter is placed under defined initial and boundary conditions, in a way intended to induce a particular event. By varying the initial and boundary conditions, one can then investigate whether, and if so how, the events induced are correlated with one another, and whether this correlation might possibly suggest the existence of a connection due to natural law. This is precisely the way in which Galileo Galilei, the founder of the modern experimental method, proceeded: he verified his famous law of falling bodies by rolling an object down an inclined plane from various starting positions.

In an experiment one cannot only confirm presumed relationships; one can also rule out wrong hypotheses. The disproof of a hypothesis is even more important than its confirmation, because only the refutation of wrong ideas finally leads to a gain in our knowledge. Even so, the mere fact that a correlation is seen between the results of measurements in an experiment is far from proving a law-like relationship. Rather, the correlation must be reproducible. Yet, every experimenter knows that in reality no experiment is reproducible in all its details. Even the starting conditions of the experiment, such as the numerical values on a measuring device, can never be reproduced exactly, because every setting and every reading of measurable values is subject to a certain error, small as this may sometimes be. This, however, is not surprising. It is a consequence of the fact that every process in the world, and therefore also the conduct of an experiment, is by its very nature a historical process.

If a physicist nonetheless asserts that an experimental result is reproducible, then he will be using the phrase in the sense that the result can be reproduced within narrow fluctuation margins. Only in cases where the processes that take place within a series of experiments do not differ too much amongst themselves can one speak of a regularity that is due to the operation of a natural law upon matter. In other words: Only when similar causes lead to similar effects can we neglect small fluctuations in the reproduction of an experimental result.

Thus, the phenomenon of reproducibility, which is of such central importance in the exact sciences, rests upon a particular form of causation. To see this more clearly, we first have to distinguish between the "principle" of causality and the "law" of causality. The principle of causality is a physical interpretation of the philosophical pinciple of sufficient reason ("*nihil fit sine causa*"), according to which nothing happens without a cause. The law of causality, in contrast, leads on the statement that the same causes always have the same effects. This in turn is tantamount to the statement that there is a regular connection between the events. The particular form of this connection is then expressed by general causal assertions, which in the natural sciences are referred to as natural laws. However, in order to get an empirical grip upon any such law the extended (or strong) form of causation must hold, according to which not only the same causes have the same effects, but also similar causes have similar effects. Only this strong form of causality ensures, despite the inherent historicity of every event, a roughly reproducible behaviour of natural phenomena.

The notion of reproducibility, which lies at the bottom of the traditional understanding of the concept of scientific law, is again an idealisation that can only be applied to a particular class of causal relationships. However, there are also law-like processes in Nature in which a different form of causality operates. In this case the causation is characterised by the assertion that *similar* causes have *different* effects.

The discovery of this kind of causation is not new. Henri Poincaré [6] already encountered this phenomenon at the beginning of the 20th century in his investigation of the mechanics of celestial bodies. Systems in which this particular form of causality operates have remarkable properties. Their dynamics are—in spite of being dominated by deterministic laws—so irregular and random-looking that one gains the impression that these systems are governed entirely by chaos. Yet, it has been found that here the laws themselves are responsible for the chaotic behaviour of these systems. This phenomenon, with its peculiar dynamics, is referred to in physics as "deterministic chaos".

The actual reason for this seemingly unruly and chaotic behaviour is to be sought in the non-linear link between cause and effect. Mathematically, a non-linear system of this kind is one in which the individual forces of interaction do not add up linearly to a total force; instead, the resultant of the forces constitutes a higher-order sum. A characteristic property of such systems is that their dynamics depend extremely sensitively upon the initial conditions (Fig. 8.2). As these systems follow exact laws, they are indeed still calculable in principle. However, in practice, their calculability is dramatically reduced, as in consequence of their non-linear behaviour the slightest error in setting the initial conditions is amplified and re–amplified in a manner resembling an avalanche. This makes long-range predictions of the dynamics of those systems practically impossible.

Disregarding for the moment their enormous physical importance, we can see clearly that chaotic systems offer an interesting model the study the phenomenon of historicity. This is because the processes that take place in such systems are neither reversible nor repeatable. They are just as unique as all other historical events. However, it is very remarkable that these systems are still governed by deterministic laws.

8.3   The World of Boundaries and Constraints

Let us now pursue in a little more depth the ideas that we have developed up to now, by focussing on the starting conditions from which a physical process sets out. Normally the law-like behaviour of a physical system is described by differential equations. As mathematical expressions of the prevailing laws the differential

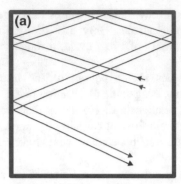 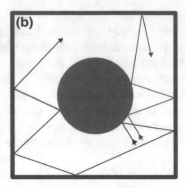

Fig. 8.2 In a system that is dominated by deterministic laws there can be two kinds of causation. The difference between them is illustrated by the movement of a particle. In physics, one uses the representation of a particle in "phase space"; this is a mathematical space based upon the position and velocity co-ordinates of a particle. Each particle can be represented by a point in phase space, so that the movement of the particle is shown by a trajectory in phase space. **a** Similar causes lead to similar effects, and the particle's trajectory in phase space is stable. Small changes in initial conditions have only correspondingly small effects upon the final state; the trajectories remain close together, and the sequence of states is reproducible. **b** Similar causes engender dissimilar effects, and the trajectory in phase space is unstable. Minute changes in the initial conditions have macroscopically visible consequences. The trajectories in phase space diverge, so that even closely similar initial states lead to the system's jumping around in an unpredictable manner, and to completely different final states. The sequence of states is irreproducible. Such a situation is given when a round obstacle is placed on a billiards table. This arrangement is known as "Sinai billiards", after the mathematician Yakov Sinai

equations do not contain any facts about the world but rather information about everything that is, in accordance with the laws, possible. The set of possible solutions is then restricted by the initial conditions from which such a system starts. The initial conditions, strictly speaking, function as a selection condition that narrows the diversity of possible solution of the equations down to those solutions describing the actual behaviour of the system.

Thus, only the initial conditions contain facts about the world and impart a concrete meaning to the otherwise abstract nature of the prevailing laws. If one were able to trace back the history of the initial conditions, perhaps even to explain them on the basis of some law together with some foregoing "pre-initial" conditions, then one would have an abstract scheme of a historical world-description (for further details see Sect. 9.3).

For a long time, considerations of this kind did not play any part in mechanical physics. This was due to the fact that the laws of mechanics are reversible and thus not associated with any direction of time. In other words: The mechanical picture of the world has no historical dimension, because all processes are symmetrical in time and, in that sense, without history. Only for this reason was it possible for the idea of "Laplace's demon" to arise in the 18th century—a demon, who is able to exploit an exact knowledge of the present state of the world, in all its details, by looking, on the basis of the laws of mechanics, with equal clarity into both the past and the future. However, as the limitations of reversible physics became increasingly visible and the principles of irreversible physics took on ever more importance, the question of initial conditions moved into the focus of physicists' interest. An impressive example is the attempt to explain the temporality of the world on the basis of the laws of mechanics and the assumption of exceptional initial conditions of the universe (Sect. 7.3).

The growing importance of initial conditions is underlined not least by the discovery of "chaotic" systems. Here, under the influence of non-linear laws, the initial conditions can even become a critical quantity insofar as the smallest disturbances and fluctuations, so-called instabilities, can reinforce themselves, setting a unique course for the physical development of a system. It is evident that his behaviour gains especial significance when the dynamics of the system allow feedback to its initial conditions. An example of such behaviour is found when the final state of a system is not stable, but in itself becomes the point of departure for a new development. In such cases the starting conditions can develop, as it were, a life of their own, because the physical processes that they set off in turn become modified.

Feedback mechanisms of this kind are important in all biological reinforcement processes. Their behaviour is summarised by technical terms like "autocatalysis" or "self-organisation". However, as in such systems the beginning of the system loses itself in the system's development, there is little purpose in continuing to use the expression "initial conditions". It would be truer to say that the initial conditions here have the character of permanent "boundary conditions" that channel the development of the system. The concept of boundary conditions is taken from physics, where it refers to the conditions that select, from among the huge number of possible processes allowed by a natural law, those processes that actually do take place in a system.

As mentioned before, in autocatalytic systems such boundary conditions can also arise on their own. This process may further lead to the optimisation and gradual migration of the system to a higher level of organisation. Seen in this way, the boundary conditions also determine the material organisation of a system. The process of self-organisation is universal in Nature and can in principle lead from simple, inert states of matter to the complex organisational forms of living matter, in which the boundary conditions take on increasingly specific characteristics. Thus, boundary conditions determine decisively the development of a self-organising system, and, since this development influences the boundary conditions themselves, the development of the boundary conditions reflects the history of the system in question. On the other hand, the concept of boundary conditions also leads into an abstract world, one that is not so easily penetrable for the non-scientist. For this reason, we will approach the topic of boundary conditions again from another direction in order to illustrate their far-reaching importance.

In this connection it is important to emphasize that boundary conditions do not only orginate by self-organisation, but that they can also be imposed upon a system from outside. Therefore, they are also referred to in physics as "constraints". This applies in particular to every experimental set-up in which Nature is forced to operate under particular boundary conditions. A famous example is the discovery of the law of falling bodies by Galileo. As already mentioned, Galileo verified the law by rolling balls down an inclined plane. However, the angle at which the plane was tilted did not have to be, and could not be, derived from mechanical laws. Rather, it had been invented and determined by Galileo.

We can see the same thing happening in a machine (Fig. 8.3). In a machine, too, special boundary conditions operate that "dictate" the working and the function of the machine. These boundary conditions are laid down in the plan for the machine's construction. The blueprint contains, among other things, a statement of what component parts the machine is built of, what material properties, shapes and arrangements these part have and what energy sources are to be used. Once all the boundary conditions are laid down, all the physical processes that take place in the machine, and which give the machine its function, are automatically determined.

The picture of a machine controlled by its boundary conditions can be transferred directly to a living organism. In analogy to a machine, the material organisation of a living being can also be considered as a complex hierarchy of physical constraints that determine the processes taking place in the organism. However, in contrast to a machine, these boundaries are not the result of a designer, but they emerged spontaneously from a long process of material self-organisation and evolution. They are shaped by evolution to steer and regulate the organic processes, in such a way as to preserve as well as to reproduce the system. In this sense, the physical boundaries determine the function of the living organism, just as the boundaries of a machine govern its working.

From this perspective, new life is breathed into the erstwhile mechanical interpretation of the organism. This idea, as is generally known, was first posited by

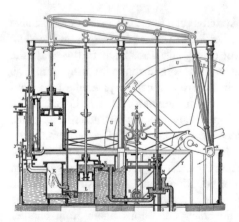

Fig. 8.3 In a machine—just as in a physics experiment—natural laws are forced to operate under defined boundary conditions ("constraints"). These in turn are a result of the design of the machine, which lays down the shapes of, and the boundaries between, its component parts. The plan and the photograph show an industrial steam engine from 1788, built by James Watt, that turns heat into mechanical energy

René Descartes, who described animals as automata without souls. In the middle of the 18th century an even more radical view was promulgated by the physician and philosopher Julien Offray de La Mettrie. He went so far as to describe in his book "L'homme machine" humans as machines functioning exclusively according to mechanical principles.

However, the mechanical theory of organisms, which later became a paradigm of biology, has repeatedly been rejected because it contradicts the popular view of Man and Nature as being organismic entities. Moreover, biological research in the 18th und 19th centuries could prove that organisms are able to reproduce and, in some cases, even to regenerate themselves. These findings, however, appeared to disprove the mechanical theory of organisms. A machine that is cut into two parts obviously cannot regenerate itself. Similarly, for a long time it was thought that a machine cannot reproduce itself.

Yet, the latter argument can no longer be maintained. In the middle of the 20th century the mathematician John von Neumann was able to demonstrate that machines which have the capacity to reproduce themselves—he called them self-reproductive automata—can very well be built. Aside from the question of how far the machine metaphor is sustainable, the focus upon boundary conditions has brought to the foreground a central aspect of living matter: On the one hand the boundary conditions encode the whole secret of living matter. On the other, the concept of boundary conditions is completely anchored in physics, and there is nothing mysterious about it.

At first glance, it seems that the boundary conditions governing a machine or a living system must be different from the physical boundary conditions that we began by describing. When we look more closely, it becomes clear that all the boundary conditions appearing in a machine or in a living organism can ultimately be traced back to physical boundary conditions. Thus the boundary conditions of a machine can, if only in a thought experiment, be described with physical exactness, by specifying the constraints operating at the boundaries between the components of the machine and thus ultimately by the properties and positions of all its atoms. The same applies for the boundary conditions in a living organism. In the latter case, the situation is in fact even simpler, because all of the "biological" boundaries can be traced back to the genome of an organism as its primary physical boundary condition. This applies irrespective of the fact that the heredity molecules remain subject to general physical constraints, such as their physical surroundings. These additional constraints complement the complex of developmental constraints that determines unambiguously the organismic processes as postulated by the doctrine of genetic determinism [4].

We have already mentioned that boundary conditions act as selection conditions. They restrict the almost unbounded number of physically possible processes to those which actually do occur in a system. The fact that boundary conditions can be very different in Nature is shown by the physical examples in Fig. 8.4. However, regardless of these differences, in all three cases a constraint applies, which restricts the freedom of movement of the system's components.

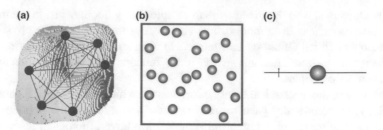

Fig. 8.4 Three examples of physical constraints. All three have in common the property that degrees of freedom of motion are restricted: **a** A rigid body (all the distances between the atoms are fixed). **b** Gas in a container (the molecules can only move inside the container). **c** Motion of a bead constrained to move on a moving wire. Constraints can be classified according to various criteria. Usually, this is done primarily by the form of their mathematical description ("holonomic" or "non-holonomic") and thereafter by whether or not they are time-independent ("scleronomic") or contain an explicit time-dependence ("rheonomic"). The constraint operating on system (**a**) is holonomic, on system (**b**) non-holonomic and on system (**c**) rheonomic

It can be generally stated that boundary conditions are responsible for the way in which a system is constructed, independently of whether it is a physical, a biological or a technical system. This is tantamount to saying that boundary conditions are carriers of information which determines the material organisation of the system. In the special case of machines and organisms the boundary conditions even encode some functional organisation.

Let us consider the living organism from this standpoint. In the development of the organism from a fertilised egg cell the information of the genome is expressed step by step in the material structure of the living organism. Since every step of the expression of the genetic material alters the physico-chemical environment of the information its "interpretation" steadily changes as well. Thus, the overall process of development is subject to internal feedback of an extremely complex nature. As the process passes through its various stages of organisation and degrees of complexity, ever newer feedback loops are formed, and this finally leads to a strong hierarchical order in the organisational structures of the organism and thus in its boundary conditions.

Ultimately, all the boundary conditions present in the final, differentiated organism are determined by its genetic information (Fig. 8.5). More precisely, the genome represents, through its material structure, the primary constraint, which is— together with its physico-chemical milieu—a necessary and sufficient condition for the existence of the organism.

The first to point out the fundamental importance of boundary conditions for an understanding of living matter was Michael Polanyi [7]. However, instead of pursuing the idea of material self-organisation, as described above, Polanyi took up an antireductionistic position in respect of boundary conditions. He considered that boundary conditions, by their nature, surpassed the explanatory potential of physics, because (in the machine as in the living organism) they carry plan- and goal-directed information. According to Polanyi, neither the laws of physics nor the

Fig. 8.5 The molecular
structure of DNA as an
example of a biological
boundary condition. **a** Side
view of the double helix.
b View down the double
helix's long axis. In principle,
DNA too could be described
as an extremely complex
physical constraint. However,
to do this one would have to
specify the exact spatial
position of each atom in the
molecule, and also the
environmental conditions to
which the DNA is exposed
(temperature, ionic strength
etc.)

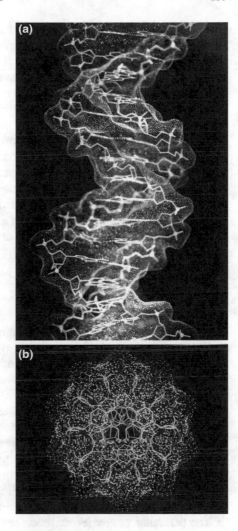

principles of statistics could ever allow us to apprehend how information-carrying
boundary conditions are selected from the unlimited variety of physically equivalent alternatives.

At the time when Polanyi published his thesis of the irreducibility of boundary
conditions, the physics of self-organisation was still in its infancy and the theoretical insights that we can appeal to today lay in the future. Nowadays we know
that Polanyi's conclusions were wrong. Modern theory of the origin and evolution
of life has shown that self-organisation can indeed proceed step by step along a path
from unspecific starting conditions toward information-bearing boundary conditions. As developmental biology and neurobiology also rest to a large extent upon
models for self-organisation, boundary conditions attain a universal significance for
an understanding of the phenomena of life [5].

More than this, we have a certain justification for the hope that the concept of boundary conditions is capable of extension toward, and application in, social systems (see Sect. 9.4). In particular, it seems reasonable that the concept of boundary conditions may be applicable to the various forms of human actions. This is because human actions are always context-dependent, that is, they are embedded in a complex hierarchy of boundary conditions and constraints. To these belong the social environment of the subject, his cultural horizons, biography, personality structure, biological constitution and so forth. All of these boundary conditions channel his activity and thus also determine his social behaviour.

Here we see in outline the significance of boundary conditions for an understanding of the real world. A theory capable of describing the development of boundary conditions might not only be the basis for a unified understanding of self-organisation processes in Nature and in society: it could also open a completely new perspective upon the phenomenon of historicity.

8.4   Outlook

The methodological approach that we have developed here, with the purpose of finding a causal-analytical explanation of historical processes, does not aim to provide a prognosis of the future course of history. Rather, it seeks an answer to the question of the true essence, the distinctive quality of historical processes, a question thrust upon us by the temporality and the uniqueness of all events. Both aspects of history, as multilayered and as complex as they may be, can indeed become research objects of within the exact sciences that set out to find a law-like explanation of reality.

It is precisely the development in the physics of complex systems that have led to a new concept of deterministic law, one that is no longer be considered to be tantamount to the concept of reproducibility. On the contrary, it has turned out that deterministic laws can be a source of unique and individual behaviour in Nature.

The particular properties of those laws are based upon a particular form of causality, according to which similar causes lead to different effects. Thus, in a non-linear system it may happen that all processes are completely irregular and unpredictable at the level of their manifestation, although at its root the system is completely subject to deterministic law. With this discovery, even the unique phenomena characteristic of historical events move into the realm of the exact sciences.

To model historical processes we can conveniently make use of the concept of initial and boundary conditions. In fact, wherever the issue of the historicity arises in the exact sciences, the initial and boundary conditions stand at the centre of the explanation. This is just as much true of the origin of the universe as for the origin of life or the origin of temporality. By looking deeper and deeper into the history of the development of initial and boundary conditions, we approach ever closer to the origin of all things.

Given this fact he astrophysicist Stephen Hawking [2] once tried to bring the chain of explanations to an end. He proposed a model of the universe without boundary conditions, one that offers an internally consistent physical explanation to meet the apparently endless quest to understand the beginning of the world. However, laws, initial conditions and boundary conditions make up the fundamental structure of scientific explanations and seem, as such, to be indispensable in science.

Even if the idea of initial and boundary conditions as sketched out here is anchored (mainly) in the natural sciences, it still does not lead to a "naturalistic" understanding of history. Rather this concept should be assigned to the so-called "structural sciences", the objects of which are the abstract structures of reality— independently of where each of these structures may come from, and of whether one encounters them in physical, biological or social systems (see Chap. 9).

Finally, human actions may have to be re-examined in the light of their social boundary conditions. This is because social boundary conditions, not only structure our life-world, but also reduce in many ways the individual's degrees of freedom. They rest like a constraint upon all our wishes, intentions and decisions. For this reason the idea that the course of human history has principally been determined by "free-will" decisions may turn out to be a mere fiction.

It may rather be phenomena of mass action that give us an understanding of the world of history. Phenomena of this kind, which may appear to us as typical expressions of complex and unpredictable human behaviour, are indeed accessible to the exact sciences. There is nothing inscrutable about them. On the contrary, they can not only be classified exactly, but also described in a detailed, mathematical way.

The models show that random and unpredictable individual behaviour can be governed by global control mechanisms, or directed by co-operative interactions. Let us consider as an example once more the reaction of a substance A with another substance B to give the product AB (see Sect. 7.2). According to the principle of microscopic reversibility, for each reaction in a particular direction ("forwards") there is also the possibility of a "reverse" reaction. Chemical equilibrium is therefore a "dynamic" equilibrium, in which the average rates of formation and decomposition are exactly equal. All fluctuations about the equilibrium value are self-regulating, following the law of mass-action: the greater the deviation from equilibrium, the greater is the rate of the counter-reaction. On average, over time, a certain, fixed concentration ratio is reached between the starting materials A and B and their reaction product AB; the value of this ratio is given by the law of mass-action, which thus regulates the inherently random behaviour of molecules.

Coherent individual behaviour can also result from "co-operative" interactions. Thus, for example, co-operative behaviour patterns always emerge where the atoms or molecules of a system have a tendency to adopt a state that is influenced by the state of their immediate neighbours (Fig. 8.6). The co-operativity can be "positive" or "negative", that is, the atoms or molecules can prefer the same or the opposite state with respect to their neighbours, so that reinforcement or dampening takes place. As in the case of the mass-action effect, this co-operativity is a statistical

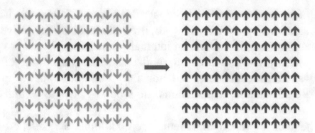

Fig. 8.6 Schematic representation of co-operative behaviour. Imagine a dynamic system consisting of elements that each can adopt either of two possible states (↑ or ↓). If there exists a coupling between adjacent elements, such that the probability of change of an element's status depends upon the degree to which its neighbouring elements have made the same change, then one speaks of a co-operative interaction. Such coupling can operate over a short or a long range, and it can be "positive" (an element tends to adopt a state similar to that of its neighbours) or "negative" (an element tends to adopt a state opposite to that of its neighbours). In this example, a nucleation seed has arisen in the centre of the randomly distributed elements (*left-hand* graphic), and positive co-operativity has induced a phase transition in the entire structure (*right-hand* graphic)

phenomenon. Positive co-operativity tends to reinforce small, random fluctuations; negative co-operativity tends to reduce them.

Co-operative interactions play an important part in both inert and living matter. The magnetisation of an iron rod, the condensation of a gas to a liquid or the crystallisation of a molten mass can all be explained by co-operative models, just as can the regulatory properties of enzyme molecules, the aggregation of virus particles and the coherent behaviour of nerve cells. At the level of biological macromolecules, co-operative interactions make up the most important physical principle, as it makes possible the rapid building-up and breaking-down of large molecular structures without compromising their stability.

Regarding the possible calculability of human actions, the most important co-operative processes are those that direct individual behaviour. Here too, there are interesting model systems for these in biology. Everyone has seen the fascinating pictures of flocks of birds, or shoals of fish, moving together in a way that changes abruptly but still with a uniformity of movement that is nothing less that the expression of co-operative behaviour.

In fact, the theory of co-operative processes is of such enormous generality that it can be applied equally to atoms, molecules, fish, birds and humans. One could imagine an absolutely comparable structure theory of historical processes, one that describes the forms of natural history and cultural history as a structural change by appeal to unified principles. For example, it is well imaginable that the course of history may one day be represented quite generally as a sequence of phase transitions that have a structure resembling those of physical and biological phase transitions, ideas with which we today describe transitions between two states of aggregation or changes in population states.

References

1. Gadamer, H.-G.: Truth and Method. Continuum, London/New York ^{2}(2004). [Original: Wahrheit und Methode, 1960]
2. Hawking, S.W.: The boundary conditions of the universe. Pontif. Acad. Sci. Scr. Varia **48**, 563–574 (1982)
3. Heidegger, M.: Being and Time. SUNY Press, Albany/New York (2010). [Original: Sein und Zeit, 1927]
4. Küppers, B.-O.: Information and the Origin of Life. MIT Press, Cambridge/Mass. (1990). [Original: Der Ursprung biologischer Information, 1986]
5. Küppers, B.-O.: Understanding complexity. In: Beckermann, A. et al. (eds.): Emergence or Reduction?, pp. 241–256. De Gryter, Berlin/New York (1992)
6. Poincaré, H.: Les Méthodes Nouvelles de la Mécanique Céleste III. Gauthier-Villars et fils, Paris (1899)
7. Polanyi, M.: Life's irreducible structure. Science **160**(3834), 1308–1312 (1968)
8. Weber, M.: Gesammelte Aufsätze zur Wissenschaftslehre. Mohr Siebeck, Tübingen ^{7}(1988)

Chapter 9
Where Is Science Going?

In the light of modern science and technology, the borders between living and
non-living matter, as well as between natural and artificial life, are increasingly
being eroded away. The most important indicator for this development is the ascent
of the structural sciences, which are designed to investigate the overarching
structures of reality. This opens up a hitherto unimagined scope for forming reality.
To do this, new means of access to knowledge are required. Can scientific information be linked up into a technology of knowledge?

© Springer International Publishing AG 2018   165
B.-O. Küppers, *The Computability of the World*, The Frontiers Collection,
https://doi.org/10.1007/978-3-319-67369-1_9

9.1   The Search for the Unity of Knowledge

In a world that is becoming increasingly complex and unmanageable, in which profound political, economic and social upheavals are taking place, more and more people are placing their hope on the help of science for finding their way and making their decisions. However, the significance of science for modern life seems to be highly ambivalent. On the one hand, science has a constitutive function in forming our understanding of human nature and of our picture of the world. On the other, science is becoming more and more specialised and compartmentalised. To the extent, however, that science is losing its broad view of reality it seems to promote the loss of contact with that same reality.

Yet the ambivalence of scientific progress does not affect the two broad streams of knowledge, the humanities and the natural sciences, in equal measure, as—according to the traditional view—they represent two fundamentally different ways of knowing reality. Thus, the humanities are first and foremost concerned with sensing the meaning of our living world. Their preferred method is that of hermeneutics, with which they set out to interpret and to understand the unique and historically evolved phenomena and their contextual meaning. The humanities claim for themselves a genuine closeness to reality, related to reality as a whole.

The natural sciences, in contrast, attempt to understand and to explain reality by its law-like behaviour. Their preferred method is an analytical research strategy based upon the principles of simplification, reduction, idealisation, abstraction and generalisation. For this reason the natural sciences frequently hear the accusation that their methodological approach necessarily leads to a loss of reality which ultimately prepares the intellectual ground for the progressive destruction of the natural habitat of human beings by man himself.

However, the discomfort with and mistrust of science is not just a product of our own age. It is well known that similar feelings motivated Goethe to pen his polemics against the mechanistically oriented natural sciences of his day. Goethe took up arms against Newtonian physics not only because he had different ideas about the nature of light, but also—and presumably much more—because he entertained a deep aversion to the mechanistic view of Nature and the looming world of technological mastery over Nature.

Despite the differences in the self-understanding and the public awareness of the humanities and the natural sciences, we will generally term both streams of knowledge as "science". Nevertheless, we must always bear in mind the significant differences that demarcate the exact sciences from the humanities, in respect both of their goals and methods of gaining knowledge.

The antagonism between the holistic and analytical views of reality which even today is still reflected in the scientific landscape has dominated the debate about the fundamental basis of knowledge for as long as can be remembered (see also Sects. 1.3 and 2.3). Philosophy of science has tried again and again to bridge this divide, by demanding, on the basis of the striven-for unity of science, the wholeness, completeness, and coherence of scientific knowledge. Only the systematic ordering of scientific knowledge, it is asserted, will lead to an objective overall view and thereby restore the unity of reality, even if its analytical dissection is a necessary prerequisite for gaining scientific knowledge.

Thus, in spite of substantive disparities between the individual schools and lines of thought, the philosophical conceptions of modern science agree in one central point, namely: that the unity of scientific knowledge is the indispensable goal of the scientific enterprise. More than this: The unity of scientific knowledge appears actually to be the prime foundation of the claim that this knowledge is valid and true. In the 19th century, when the idealistic philosophy was at its height, the idea that the absolute is the highest principle unifying our knowledge ran rampant (see Chap. 1).

Even though the idealistic aberrations have been strongly fought against by the exact sciences there will hardly be any scientist who can resist the fascination exercised by the idea of the unity of knowledge. However, it must also be admitted that the conceptions of this unity, so often evoked, are still relatively blurred. Most commonly, the idea of unity is associated with the search for the unity of science. Yet here too it remains unclear how the goals, methods and areas of application of the various scientific disciplines can be subsumed under the form of unity, and what we then would actually be supposed to understand by the term "unity".

In the face of these difficulties, we recognise in the idea of unity merely a guiding motive of modern science, one that apparently has a regulatory function in acquiring knowledge of reality; nonetheless, its principal features are now beginning to emerge little by little, as scientific knowledge progresses.

Another point needs to be clarified: The problem of the unity of science is not just one involving the relationships between the individual disciplines—that is, an *inter*disciplinary problem; rather, it is also a *trans*disciplinary problem, in that the borderlines between the various disciplines are crossed, with the result that these borderlines may one day abolished altogether by the idea of the unity of science.

Thus it is more than a mere quibble over words when in the coming paragraphs we make a distinction between the "unity of *science*" and the "unity of the *sciences*". In the former (unity of *science*), the idea of unity would appear to dissolve all the differences between the scientific disciplines. According to this idea, science attains its highest level of perfection in the erasing of all borderlines between the various disciplines. In the latter (unity of *sciences*), the individual scientific disciplines seem to have retained their autonomy. The various disciplines then continue to embody different routes of access to reality. Nonetheless, they are linked by a unity in which their various models of reality can be put into common service as analogy models.

In the following pages we shall address the idea of the unity of the sciences in the sense just described. An important touchstone for this will be the question of whether, and if so in what form, there can be any mutual approach of the exact sciences and the humanities. Even if the unity of the sciences cannot at present be verified, at least the relationship between the exact sciences and the humanities can be expected to act as a good indicator of the staying-power of this idea.

In principle, the issue can be approached from two sides. On the one hand, we can perform a logical analysis of the existing concepts and theories of science and attempt to test the idea of unity from a metatheoretical standpoint. On the other, we can restrict ourselves to the description of the general trends in the development of contemporary science that suggest a possible convergence of the sciences in respect of their subject-matter, their goals and their methods.

The merely descriptive procedure has the advantage, compared with the metatheoretical approach, that intricate questions of the philosophy of science, such as the inter- and intratheoretical reduction of scientific theories, can be avoided. Instead, one considers only those features of the development of the sciences that can, so to speak, be seen at a glance. Its disadvantage consists in the fact that this procedure does not include an explication of the concept of unity.

Here we set out on the latter path and, first of all, take a brief look at the roots of the dichotomous understanding of science that predominates today. We shall then continue by eliciting the aspects that determine the antagonism between the humanities and the exact sciences, in particular the natural sciences. In the final part of our analysis we present a new type of science that adopts a bridging position between all scientific disciplines and which, by doing so, may be in a position to mediate the unity of the sciences.

9.2   The General and the Particular

Let us first consider the splitting of the scientific landscape into the exact sciences and the humanities. The assertion that there is a convergence between these two major currents of scientific research is at first a mere working hypothesis. Its truth is by no means evident, especially as it appears to contradict quite fundamentally the present-day understanding of science.

One only needs to mention the famous essay "The Two Cultures" by Sir Charles Snow [17] which, published in the 1960s, attracted much public attention. This essay put forward the general thesis that in Western society the scientific-technological intelligentsia and the literary-artistic intelligentsia regarded each other with an attitude that varied from irreconcilability to downright hostility, because these social groups—according to Snow—are representatives of two deep-rootedly different cultures of knowledge. Thus, for anyone who basically subscribes to this thesis, there can be no talk of any mutual approach, to say nothing of any unity, of these two cultures.

The cultural antagonism diagnosed by Snow is in step with the prevalent dichotomy of the scientific disciplines into natural sciences and humanities. According to this view, the natural sciences are dedicated to the ordering of reality as expressed in the laws and regularities of Nature. Their goal is to uncover the causal relationships between natural events and to explain them by appeal to natural law. Their methodological approach is to ignore the specificities of reality and to replace the objects of knowledge by idealised and mensurable quantities.

In contrast to this, the humanities claim to focus upon the intellectual and social order of reality, as expressed in state and society, in jurisprudence and economy. They provide an interpretation of reality by means of language, religion, myth and art. Their method, as practised above all by the adherents of the hermeneutic doctrines, is aimed at an interpretive understanding of reality and not at an explaining it by reference to laws.

The roots of the dualistic concept of science can be traced back to a philosophical viewpoint which was adopted by Wilhelm Dilthey [5] at the end of the 19th century and which stresses the autonomy of human's intellectual life. At the centre of this philosophy is the idea of understanding mental reality "from the inside". By this, Dilthey meant the "internal perception" of those order-structures that lie behind of our intellectual life and which have made their imprint upon the symbol- and rule-systems of human language and actions, also including creative acts. Dilthey argued that in contrast to the natural science which find their objects of cognition independently of any expression of human life and which thus can consider the order of Nature from the outside, the humanities are forced to explain and to understand the essence of the intellectual life intrinsically. He was convinced that the internal perception, the retracing und reliving of the mental processes is the only way, in which the objects of cognition in the humanities come about. On the basis of these considerations he reached the conclusion that mental phenomena make up a separate, independent area of experience over against material phenomena.

However, in contrast to Dilthey his contemporary Wilhelm Windelband saw the need to view the antagonism of the natural sciences and the humanities as a relative matter. Nature and mind seemed to him to stand in a close relationship of mutual dependence—much too close to permit them to be regarded as the foundation-stones of two completely different forms of science. Moreover, Windelband was not satisfied with the postulate that the objects of the humanities were constituted solely by internal perception. It also disturbed him that important, experience-based sciences such as psychology did not fit into the bipartite schema that Dilthey had set up. According to its objects of knowledge, psychology belongs to the humanities, but its methods are those of the natural sciences.

Windelband believed that one could resolve the "incongruence" of the practical and formal classification of sciences by replacing the antithetical pair "Nature/ Mind" by the antithetical pair "Nature/History". He was of course clear that the displacement of the perspective from mental to historical occurrences did not resolve the dualistic view of the sciences. However, he believed that the underlying dichotomy would gain in incisiveness, because its conceptual and systematic positioning would now appear with far greater clarity.

A particularly illuminating statement of Windelband's understanding of science can be found in his lecture about "History and Natural Science" [20]. In this lecture Windelband attempted to justify the division of the sciences into natural and historical sciences, as a manifestation of our understanding of the world that rests upon two basic elements: "law" and "event". According to Windelband, the natural sciences search for the "general", in the form of universally valid laws, while the goal of knowledge in the humanities is the unique event, the "particular", in its historically determined form. The one set of disciplines considers the "form", which always remains the same, and the other the unique and intrinsically given "content" of the real event. The one is concerned with what always *is*, the other with what once *was*.

To characterise this dichotomy, Windelband introduced the technical terms "nomothetic" and "idiographic", which are tantamount to "law-like generalisation" and "individualising description". Later, Heinrich Rickert [16] sharpened up the distinction between the sciences of laws and the sciences of events, by re-interpreting the latter as a general "cultural science", which is primarily concerned with the determinations and assignments of values and meaning in our cultural world.

The philosophical problem which Windelband addressed in his lecture is as old as philosophy itself. It is the question of the relationship between the general and the particular: Can the particular ever become the object of a science, which has the general as the goal of its knowledge? Windelband's answer to this question was a clear "no". Indeed, this answer seems to be inevitable when one follows the tradition of thought, reaching back to antiquity, according to which the general and the particular are two incommensurable elements of our intellectual world.

Aristotle, for example, defined the general as that which, by its very nature, can be immanent in several individual things at once. In contrast to the general, Aristotle claimed, no positive characteristics can be attributed to the particular; it can only be specified negatively, as that which cannot be narrowed down by general concepts. For this reason, Aristotle emphasised, no knowledge of the particular is possible, but only perception of the particular by our senses.

More than that, in contrast to the particular, the general is not only an object of knowledge, but it is also the prerequisite for knowledge. Only if the manifoldness of phenomena can be traced back to a unity does an understanding of that manifoldness become possible. The unity and the generality of phenomena in turn manifest themselves as a rule, a principle or a law.

If one accepts the Aristotelian view according to which it is impossible to grasp the particular with the aid of general concepts then any law-based specification of the particular is excluded a priori. Within the framework of this logical thought, the relationship between the general and the particular emerges as a cardinal problem of scientific thinking, one for which there is apparently no solution. Thus, the dualistic conception of the sciences, such as Windelband conceived of, appears to be inevitable.

Nevertheless, there is still a way in which the *aporia* of the unbridgeable contrast between the particular and the general may be overcome (see also Sect. 5.3). This way forward was proposed by the philosopher Ernst Cassirer. Unlike Windelband and Rickert, Cassirer rejected the strong historisation of the humanities (or "cultural sciences"). Instead he gave in his major work "The Philosophy of Symbolic Forms" a prominent place to the formal and binding character of all sciences [3]. Especially, Cassirer provided a foundation for the nature of the particular, based not upon historical uniqueness, but rather upon the unique position that the particular adopts in its complex network of relationships to reality. In other words: For Cassirer, the particular is not particular because it eludes any general determination, but rather because it does precisely the opposite: it becomes particular by entering into more and more circles of the general and in that way, it falls more and more into a *specific* relationship to the reality.

The "moulding" of the particular in the context of its general relationship circles brings out the specific attributes of the particular and makes it possible to describe it by using formal and general concepts. In his earlier, systematic treatise on "Substance and Function", Cassirer asserts: "There is no reason to believe that any concrete content needs to lose its particularity or its clarity as soon as it is lined up with other similar content in various orders and contexts and thus made 'conceptual' and formed. The opposite is closer to the truth: the further this forming progresses and the more relationship circles are entered into, the more sharply defined and specific it becomes" [2, p. 298].

According to this picture, the particular gradually "crystallises" out of the network of its general determinants. In that way the particular is positively determined —through its functional relationships to the general. In contrast, Aristotle excludes precisely this possibility of a positive determination of the particular based upon the general. He characterises the particular only in a negative way, namely by demarcating it from the general.

Cassirer had rightly insisted that any talk of the particular, the individual or the unique makes little sense in the science of history unless we succeed to find a general approach to the phenomena of history. As he argued, we can obviously only pursue science when we abstract from the specific—that is, individual and unique— features of historical events. Even the mere naming of the objects of a scientific enquiry presupposes that we can assign conceptual properties with a certain degree of generality to these entities. Yet in the moment when we reach an intersubjective understanding of the individual and the unique, it rises to the position of becoming an object associated with general properties and is thus, at least in principle, accessible to explanatory and descriptive science employing a law-based approach.

Even Windelband had had to admit this. At the end of his lecture about "History and Natural Science" he pointed out that a science of the individual and the unique would probably not be possible, without recourse to the conceptual categories of comparison, abstraction and generalisation. However, these categories belong to the characteristic features of a nomothetic science. Our discussion in Chap. 8 on the possibilities of a law-like understanding of history must be seen precisely in the light of these insights.

From this point of view the different ideas of Aristotle and Cassirer concerning the relationship between the general and the particular become quite understandable. More than that, it now becomes clear that there is no real contradiction at all. Aristotle argued from the position of the logician who only distinguishes between true and false statements. These are independent of time. They refer in this case exclusively to the conceptional definition of the general and the particular. In contrast, Cassirer was seeking the *genesis* of the particular. He considered only the logic according to which the particular achieves its unique properties. In the limiting case, when the particular has become manifest as such, the Aristotelian view becomes valid without any restrictions. In other words, the law-like understanding of the particular as such is completely different from the law-like understanding of the genesis of the particular. Insofar, the different views of Aristotel and Cassirer are entirely compatible.

The procedure we proposed in Chaps. 5 and 8 for a scientific approach to the concept of semantics as well as to the concept of historicity follows precisely Cassiers' figure of thought. Thus, we were *not* looking at the particular as such but rather at its genesis in the context of general determinants. For this reason we started with simple questions and models, hiding to begin with the full content of the concepts of semantics and historicity. However, this is nothing other than the well-known strategy of the exact sciences when they apply their instruments of simplification, reduction and abstraction in order to approach the complex and unique phenomena of our world.

9.3   Scientific Explanations

The recognition of the fact that the gap between the general and the particular is not unbridgeable is an important step on the way toward the unity of the sciences. In fact, present-day developments in science appear to confirm this to the fullest extent. There has been a blurring and overlapping of the borders between disciplines, to a breadth that calls the dualistic understanding of science fundamentally into question. The trend toward inter- and transdisciplinary research is characterised above all by the increasing degree to which even the historically evolved, complex and unique structures of reality are traceable to general laws, principles and rules.

In this way, a development is continuing that reached its first peak in the 19th century, when Charles Darwin succeeded in tracing the unique history of development of life back to a natural principle. Windelband was also highly alert to the possibility that the complex events taking place in the natural world might one day find an explanation within the framework of the law-based sciences. However, restricting his argument, he pointed out that for the law-based sciences the goal of knowledge does not comprise the unique manifestations of Nature; rather, also in this case it is only on the basis of its regular behaviour that natural history can be investigated and described in a law-based way. In contrast, the (in the strict sense) historic traits of living matter, which are manifest in particular developmental characteristics of organisms, appear to be closed to any investigation within the law-based sciences. They would, according to Windelband, always remain typical objects of the historical sciences, the goal of which is the accurate, detailed "description" of the objects of investigation. Thus, Windelband regarded biology, which includes a major part of "natural history", as a hybrid of two differing kinds of science, of nomothetic and ideographic sciences, which however complement one another in the aims of their search for knowledge.

To understand Windelband's argument in detail, we must look more closely at the structure of scientific explanations. To do this, we can best consider explanations in physics, as these are paradigmatic for law-based explanations in science.

Moreover, they have a simple structure: they rest upon only two elements: natural laws and initial conditions (see also the discussion in Sect. 8.3). The latter describe the exact physical circumstances—the "conditional" complex—under which the event to be explained occurred or will occur. These are the conditions that the physicist either chooses freely in designing an experiment, and can set up, or that he finds as a natural "given".

The prototype of a physical explanation is the explanation of planetary motion. In this example the general law that determines the regularity of the planets' movement is Newton's law of gravity, whereas the initial conditions are given by the positions and velocities of the planets at a particular moment. Within Newtonian theory, the subsequent trajectories, along which the planets move about their central star, can then—with the aid of the law of gravity—be calculated exactly from the initial values.

In the past, the centre of attention in physics was occupied largely by the law-like principles that determine the course of natural events. The initial conditions, in contrast, received relatively little attention, even though these are just as indispensable for the law-based explanation of Nature as the natural laws themselves. In fact, viewed formally, a law-based explanation consists precisely in deriving the event to be explained, with the help of natural laws, from the initial conditions. Naturally, such an explanation can only be obtained in a logically strict sense if the laws in question are deterministic ones. In the case of statistical laws, such as those that play an important part in microscale physics, the occurrence of the event to be explained can only be inferred with a certain probability.

Accordingly, the philosophy of science distinguishes between "deductive-nomological" and "inductive-statistical" explanations. Furthermore, the general structure of scientific explanation, sketched so far, allows it to take into account different types of scientific explanations. Thus, one distinguishes between causal, genetic and dispositional explanations; between *a priori* and *a posteriori* explanations; between retrodictions, prognoses and so on.

The standard model of a scientific explanation is however the deductive-nomological explanation, and its paradigmatic applications is the case of planetary motion. It has been investigated in detail by the philosophers of science Carl Gustav Hempel and Paul Oppenheim [8]. This model is also termed the "subsumption model" or "covering-law model" of explanation. Both terms are intended to express the fact that within a scientific explanation the event to be explained, the so called "explanandum", is attributed to the action of general laws.

According to this, even the nomothetic sciences are concerned with individual events. However, these are events that can be subsumed, in the sense set out above, under general laws. In contrast, historical events are clearly different in nature, as they apparently evade subsumption under general laws. Such events reveal the peculiarity of historicity that is usually referred to as "contingency". This in itself ambiguous term here refers to events that are neither random nor necessary. In other words: the notion of historical contingency characterises events that can occur, because they are possible, but that need not necessarily occur.

Admittedly, one might ask whether historical events could one day be explicable on a law-like basis, although this would annul the contingency of such events. This question was also raised by Windelband, and he, too, analyzed this question on the basis of the subsumption model of explanation. However, he focussed not upon the possible laws, governing the historical events, but upon the initial conditions from which a historical process starts. This analysis led him to the following result: The initial conditions are a manifestation the factial world. They include in particular those facts, the conditional complex, that induce a particular historical event. According to the subsumption model, however, the initial conditions cannot themselves be explained, because they appear in the explanation as prior conditions for the events to be explained. On the other hand, factual statements about reality cannot be inferred from universal laws alone, because general laws—unlike concrete events—are not restricted in either space or time, and thus they are unable to imprint any event-structure upon reality. The laws certainly provide a connection between events, but they do not themselves have the character of events.

We can conclude from this, in agreement with Windelband, that "law" and "event" are two incommensurable items of the explanation of the world. This is, according to Windelband, the proof that the idiographic sciences can never be absorbed into the nomothetic sciences, but must always remain an independent form of science. However, Windelband still suggested in his analysis that he could imagine a chain of interleaving explanations in which the initial conditions themselves would become the objects of explanation. Nevertheless, he immediately moderated his conclusion by pointing out that the initial conditions could not be explained away in this manner, as every law-like explanation of initial conditions presupposes in turn a preceding set of initial conditions, and so forth.

Windelband's central argument seems to be sound: The event character of the world, and thus its historicity, is a fundamental aspect of our reality and this fact cannot be abrogated thorough any kind of act of law-like explanation. One might object that no one has any intention of eliminating the historical dimension of world events; rather, the question at stake is exclusively the question of whether and to what extent the course of history can be captured in general laws. Windelband had in fact answered this question in the affirmative, by bringing the explanation of initial conditions at least into the realm of the possible.

Law-like explanations of historical processes do indeed appear possible if one treats the initial conditions of an event as the object of explanation and repeats this step of explanation, under the assumption of fresh initial conditions, again and again. By explaining each set of initial conditions with the help of earlier initial conditions, we obtain increasingly profound insight into the history of the initial conditions and thus into the reality of history. It is precisely this explanatory strategy (known in the philosophy of science as "genetic" explanation) that we follow in the exact sciences when we are trying to reconstruct historical processes such as the origin and evolution of life by appeal to the law-like behaviour of matter [9].

Even those aspects of historical events that, as emphasised by Rickert, are bound to values have today become the object of nomothetic explanation. Thus, for example, the self-organisation and evolution of life has been explained by reference to a physical "value principle" which governs the transition from non-living to living matter [6]. The models show that already in the history of Nature—albeit at an immeasurably lower level of complexity than in that of society—"event" and "value" are indissolubly linked one to the other. However, the naturalistic value concept has a fundamentally relative character, because the values inherent to Nature depend upon the perpetually changing conditions of natural order. Besides this the concept of value has here no moral dimension.

We can now draw a first conclusion: The borders that Windelband, Rickert and others set out with their dualistic conception of science seem today to be undergoing a process of gradual dissolution. The central elements of the ideographic sciences, such as individuality, uniqueness, value and the like, no longer lie outside the area of applicability of the law-based sciences, and are increasingly becoming the object of nomothetic explanations. This development was already palpable in early studies, in which attempt was made to explain the historically evolved structures of language, the historical development of economic systems and the like by appeal to law-like behaviour.

In view of the enormous progress that has been made in understanding complex systems, it has become impossible to duck the question: Is it possible (and if so, how) to construct a theory of historical processes that is based closely upon the nomothetic method of the exact sciences (see Chap. 8)? Naturally, this question exceeds the scope of present-day scientific knowledge. However, it is not as far removed from reality as it may first seem. Introducing historical explanations into the law-based sciences does not reduce the differences between the natural history and human history to a contourless picture. Nonetheless, it is conceivable that the two forms of historical development might show unified features, ones that could be described within the framework of a general theory of structural changes in history. In precisely this sense we may speak of a possible unity of natural and cultural history—a unity that could consist of common structures of development. Likewise, the cultural- and social-anthropological studies of Claude Lévi-Strauss [11] could be interpreted in such a way that the transition from Nature to human culture is seen as gradual, allowing the two aspects of the course of history to be described by unified structures.

If we review the arguments presented above, then it becomes clear that the sharp distinction between natural history and human history cannot be sustained. Instead, we are compelled to assent to the verdict of the mathematician and philosopher Richard von Mises: "The study of Nature on the one side, and that of history and human affairs on the other, are the two hemispheres of the 'globus intellectualis'. Subjectively, we feel that we can distinguish between them, but that makes it all the harder to draw up exact borders between them, and on closer examination the contradistinction between them seems to disappear completely" [15, p. 313].

9.4 The Ascent of Structural Sciences

Our analysis of the scientific landscape has made it clear that the borderlines between the traditional disciplines are beginning to dissolve. This development is largely due to the fact that the complex and historically evolved structures of reality are moving increasingly into the focus of the law-based sciences. A prime example of this is the concept of boundary conditions, described in Chap. 8, which contains the key to a general understanding of historical processes. However, the concept of boundary conditions is only so far-reaching because it expresses a universal and highly abstract idea of historicity, one that is in the first place free from any references to real history.

The science of abstract structures of this kind I have referred to as "structural science" [10]. This term is a translation of the German word "Strukturwissenschaft" which, to my knowledge, was first used in the above sense by the physicist Carl Friedrich von Weizsäcker [18]. The distinguishing feature of this type of science is that—unlike the natural sciences and the humanities—it deals with the over-arching structures of reality, independent of the question where these structures actually occur, whether they are found in natural or artificial, living or non-living systems. Owing to their high degree of abstraction, the structural sciences include a priori the entire realm of reality as the area of their applicability. In this way they have an authentic access to the superordinate principles that structure the complexity and which we characterise by using generic terms such as system, regulation, information, organisation, competition, co-operation, network and so forth.

Some structural sciences can already be called classical. To these belong cybernetics, game theory, information theory and systems theory. New disciplines such as the theory of self-organisation, synergetics, network theory and many others complement the present-day spectrum of structural sciences (Table 9.1). The concept of boundary conditions, which we set our sights on above, may also one day develop into a self-contained structural science, one that can be applied to all historical processes. To give this future science an appropriate name I have proposed to call it "peranetics", built from the Greek word "peras" (border, edge).

The ascent of the structural sciences began more than half a century ago. Among their founding fathers were Norbert Wiener (cybernetics), John von Neumann and Oskar Morgenstern (game theory), Claude Shannon (information theory) and Ludwig von Bertalanffy (systems theory). The archetype of a structural science is mathematics. For this reason one might be tempted to see in the structural sciences that form of science that is usually referred to as "formal science". However, the term "formal science" has for a long time had a well-worn meaning that points in quite a different direction from that of the term "structural science". The formal sciences—some authors refer to them as "ideal sciences"—have traditionally been regarded as counterparts of the "real sciences". In contradistinction to these, the formal sciences are based upon knowledge that is independent of experience, such as that expressed in the true statements of formal logic. The formal sciences have,

Table 9.1 The structural sciences provide the foundation of both the natural sciences and the humanities. The objects of their attention are the general structures of reality, which they attempt to represent by sole use of mathematical concepts and symbols and their transformations. The laws of the structural sciences are derived from empirical laws by eliminating all descriptive and empirical constants. The "structural laws" thus obtained possess the same syntactic structure, or logical form, as do the laws from which they are derived. However, they can now be applied to more than just one area of reality

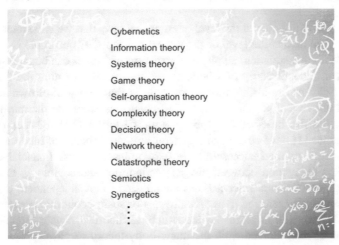

Cybernetics
Information theory
Systems theory
Game theory
Self-organisation theory
Complexity theory
Decision theory
Network theory
Catastrophe theory
Semiotics
Synergetics
⋮

so to speak, a life on their own, because their development is independent of the reality surrounding us.

However, unlike the formal sciences, the structural sciences cannot be demarcated from the real sciences. On the contrary: the structural sciences constitute the abstract scaffolding on which the real sciences are based. This is precisely the reason why the structural sciences take on a bridging function between the empirical sciences. This bridging function is also reflected in their methodology, which ignores the different forms in which reality is manifested and seeks to replace the concepts, with which we usually describe reality, by abstract concepts and symbols. All the entities, states and processes of reality are thus first treated as abstract structures, without any reference to their respective real forms of expression.

Nevertheless, and this is the actual basis of the difference between structural and formal science, the laws of the structural sciences are derived from reality. We may call these laws "structural laws". They have the same syntactic structure, the same logical form, as the real laws from which they were abstracted. Yet the objects to which such structural laws can be applied cover many different areas. Adopting an expression of the mathematician Hermann Weyl, we might also say that the structural laws represent something like a "logical casting mould for possible sciences" [19, p. 42].

As regards the question of the unity of science, the most important structural similarities are those that apply to laws in different branches of science. A simple example is the structural similarity between the physical law that describes the pressure–volume relationship of an ideal gas at constant temperature (the Boyle–Mariotte law), and the micro-economic "law" of supply and demand. According to the well-known rules of the market, a reduction in the price of a particular commodity will lead to an increase in the demand for it, and an increase in its price will conversely lead to a decreased demand. The product of demand and price is in this case as constant as the product of volume and pressure of an ideal gas.

However, it is not only at the level of empirical laws that we encounter structural similarities, but also at the level of theoretical concepts. An example is the Fourier equation. With this equation, all physical transport processes can be represented in a unified form, independently of whether they are those governing the transport of heat, electricity or liquids. A further example is the theory of Brownian motion, mentioned above (see Sects. 6.2 and 7.2), which describes the thermal motion of molecules. On the basis of this theory Robert Merton [14], Fischer Black and Myron Scholes [1] have developed a general structural theory of random processes which has been applied successfully in economics to assess financial derivative markets. The interaction between physics, economics and sociology, mediated by the methods of structural sciences, has already led to an independent direction of research that is now referred to as "the physics of socio-economic systems". At the centre of this we find in particular the structural theories of phase transitions and mass-action effects, derived from the many-body theory in physics.

Today the structural sciences are a powerful instrument for exploration of the complex structures of reality. They already link up large areas of natural science, economics and the humanities. Without the integrative function of the structural sciences, progress in research into complex systems would not only stagnate, but also impossible in many areas. This is seen especially clearly in modern biology, the theoretical foundations of which are based largely upon the structural sciences as information theory, game theory, network theory and many others. For economics too, the structural sciences have progressed to being indispensable instruments of its theoretical foundation. Alongside game theory, the origin of which lay essentially in addressing economic problems, these instruments include decision theory, the theory of self-organisation, chaos theory and the theory of fractal geometry—all of which are gaining increasing recognition in economics (see for example [13]).

Last, but not least, thinking in structural categories has—by way of semiotics and the structuralism of linguistics—also gained a firm foothold in the humanities and, from there moved on to other areas such as anthropology, sociology and psychoanalysis. Sociology has progressed to treading out its own paths, with the hitherto most comprehensive attempt to provide an exact foundation of social systems in recourse to the "functional systems theory" conceived by Niklas Luhmann [12].

Even in mathematics, the archetype of all structural sciences, meta-structural approaches have been developed with the intention of merging the numerous mathematical sub-disciplines by way of abstract hyperstructures and thus unifying them. The field was led for a long time by the French group of mathematicians known as "Bourbaki", which pursued the goal of placing all the principal areas of mathematics upon a foundation of three basal hyperstructures, which they termed mother-structures ("structures-mères").[1] These and other attempts to provide a foundation for mathematics show that even within mathematics itself the idea of unity is a driving force in the progress of discovery.

The most important instrument of the structural sciences is the computer. Computer-aided processing of huge quantities of data, accompanied by the possibilities offered by computer graphics and computer animation, today provides an indispensable basis for research into complex phenomena. In this way, computer-aided research into reality has established itself alongside the experimental methods of the natural sciences as a new path towards discovery.

The fact that computer-based simulation in natural science is increasingly replacing the traditional laboratory experiment is a striking illustration of the way in which the methods of the structural sciences are already penetrating deeply into the fabric of the traditional sciences. The question of whether virtual experience by computer simulation leads to a virtual world, one that corresponds to only a limited extent with the experienceable world, or whether it is only an extremely effective application of traditional experimental methods, can at present not be answered finally. The most appropriate comparison is probably between the use of the computer and the use of the microscope. Just as the microscope led humanity to grasp a new realm of reality, one that is not accessible through our natural perceptive abilities, computer simulation grants us insights into complex contexts that are not accessible to us through our traditional analytical tools. A wonderful example of this is the deep insight which we have gained into the structure and properties of the Mandelbrot set with the help of powerful computers (see Sect. 6.3).

9.5   Towards a Technology of Knowledge

Although the structural sciences have by now taken on an enormous significance, they have themselves remained largely uninvestigated by the philosophy of science. Yet, there is no question that the structural sciences, on the basis of their high degree of abstraction, are in a good position to create a bridgehead between the natural sciences and the humanities. Insofar as the structural sciences are desigened to explore the general and overarching structures of reality, they do not recognise any borderlines between the objects of their research. On the contrary: they transgress the traditional boundaries of the real sciences, even if they can ultimately be applied to the real world with its manifold boundaries.

It is the structural sciences that present a unified picture of reality, one that embraces all the forms of scientific knowledge. However, it may seem paradoxical

---

[1]Nikolas Bourbaki is the pseudonym of a group of mathematicians (with continually changing membership), who since 1934 have been working on a mathematical textbook, the "Éléments de mathématique" (see [4]).

that it is precisely the multifaceted science of complex phenomena that leads back
to the unity of knowledge and thereby to the unity of reality as well.

The structural sciences are occasionally referred to as "computational sciences",
as their most importance methodological tool is the computer. In fact, only the
avalanche in the development of computer technology and the vast increase in
computer capacity and power that accompanied it have given the structural sciences
their wide areas of application, stretching from information and communication
technologies through bionics, robotics and neurotechnology to computer linguistics,
socionics and economics.

The pure and applied structural sciences have by now advanced to being the
leading sciences of the 21st century. Without them, the complex phenomena of
Nature and of society would not be understandable in scientific categories. Only
with their help will we be able to gain control over the complex reality in which we
live, as the structural science will open up the way to new technologies, above all
the technology of knowledge. Such a technology is necessary, if we wish to locate,
within the sheer innumerable cornucopia of human knowledge, that knowledge
which is required for the solution of a given task or problem. This knowledge may
sometimes have to be gleaned and brought together from areas of knowledge that
lie far apart.

The need for a technology of knowledge can be clearly illustrated by using an
example from economics. It is especially economics, with its extremely complex
processes, that lacks simple algorithms for the scientific management of complexity. Economic processes are largely unpredictable, because they depend upon
social interactions involving usually vast numbers of participants.

Even if the collective actions of humans can today be predicted with increasing
accuracy, so many (partly unknown) factors are jointly responsible for the course of
economic processes that predictions can at best only be made for minute excerpts of
the overall complexity. Moreover, even deterministic processes, if they follow
non-linear laws (which in economics is frequently the case) can show chaotic—that
is, completely incalculable—behaviour. For that reason, one can with some certainty rule out the likelihood that economics will ever come into possession of
scientifically well-founded theories that allow predictions to be made that are
comparable with those of physics.

Most economists therefore regard their discipline more as a social science than
an exact, law-based science. It is also understandable that they criticise those theoretical economists who attempt to describe the extremely complex processes of
economies by using comparatively simple formalisms. Ideas like that, it is objected,
would hold up to us a picture of reality that pretends to be exact, but in fact is far
removed from the truth.

Even if the possibility of exact predictions will presumably remain an unfulfilled
dream of economics, there is still no objection in principle to the attempt to place
complex economic processes upon a scientific footing. Other scientific disciplines,
too, such as biology, are concerned with extremely complex phenomena. Likewise,
processes in natural evolution can only be predicted in rare cases, because the
course of events is strongly directed by incalculable influences, such as chance.

Nevertheless, evolution research has afforded us profound insights into the mechanisms of evolution.

Perhaps this could become a pattern for economics to follow. Even in economics, the goal cannot be to predict complex economic processes in all their details. Rather, attempt is made to understand the structural basis for such processes. Many insights can contribute to this by providing a general understanding of the dynamics of economic systems and the mechanisms of the generation of information, of structural changes, mass-action, co-operation, competition and the like. Under favourable conditions these may even make possible pattern predictions ("Mustervorhersagen"), to adopt an expression coined by the economist Friedrich August von Hayek [7]. Such exemplary predictions, which take account of the restrictedness of information and knowledge, may not allow prognoses of individual events, but they can provide a template for comparable events and as such may have an explanatory value.

In spite of the limitations that we have described, it is still possible to develop scientifically based models for the economy. This is documented by numerous concepts, which range from evolutionary approaches in competition theory through diffusion models for finance markets to models for behavioural economics. Even though theoretical economics has developed many of its approaches by reference to physical theories, it is nonetheless not a branch of physics, but rather a typical example of an applied structural science. This explains the fact that economic processes are based on events in society which in turn can only be described by general structural theories. In this respect, for example, game-theoretical models have attained especial importance in economics.

This approach to economics can be described as "structural economics". The development of such a kind of economics, which includes the entire gamut of the concepts of structural sciences, and in which all structural sciences are enmeshed, remains a task for the future. However, even if such a structural economics were one day to come about, it would not alter the fact that the concepts of the structural sciences never allow more than an overall understanding of the general aspects of economic processes.

Clearly, theoretical economics is confronted here with the fundamental problem that runs like a common thread through this book. It concerns the question of how an individual, situation-specific event can be represented by means of general concepts. The solution to this problem can only consist in a new type of scientific method, one that combines and weights general concepts in a specific way such that the economic event is more or less precisely reflected in it. This procedure is very similar to the idea we have proposed for a scientific approach to the semantic aspect of information (see Sect. 5.3). Thus it is not an issue of developing new concepts, ones that are as general as possible and cover a large area of economic reality. Rather, it is the task of developing novel "applications" of concepts that are already available. This is best referred to as "technology of knowledge".

The technology aimed at here is a sort of "knowledge engineering", with the help of which the elements of our knowledge can be assembled into a new type of knowledge, tailor-made for the solution of complex problems. Its task is to activate the innovation potential of the sciences, by putting together the requisite areas of knowledge in an interdisciplinary way and combining them to generate a solution to a particular problem. Only in conjunction with a "knowledge engineering" of this kind can the possibilities that science and technology offer us be maximally exploited to help us in mastering the challenges of the modern world.

Strictly speaking, the technology of knowledge is an applied structural science, as it makes use of higher-order structures of knowledge to place a particular piece of knowledge, along with its possible applications, in a broader context and to bring it to fruition for the solution of problems in complex systems. For this purpose, it must be able to filter out and to bundle the knowledge, which may be spread out over several disciplines, that is relevant for attacking the manifold problems of our time.

One might also characterise the technology of knowledge as a meta-knowledge that provides knowledge about knowledge. This kind of over-arching knowledge is indispensable—not least when we cross boundaries into completely new dimensions of human knowledge and activity. Advanced societies will increasingly often be faced with such border-crossings, and knowledge engineering will be the key technology with which human knowledge can be managed and monitored. Without it, knowledge-based society, the raw material of which is knowledge, will not be viable.

Even the mere access to knowledge already presupposes an efficient technology. This is related *inter alia* to the fact that all the time, world-wide, a gigantic quantity of information is being produced, in which relevant information can only be located with the help of refined searching technologies. Even the knowledge already present runs the perpetual risk of being inundated in the ever-rising tide of knowledge.

Here, knowledge technology is confronted not only by a quantitative problem, but also by a qualitative one—a problem that is of a completely new kind in science or technology. The challenge is to detect and locate precisely (within the sheer endless mass of information available) the specific information that, in a particular connection, can be processed into meaningful knowledge. A disorganised mass of information, in the form of unstructured data, does not yet constitute knowledge: at best, it contains potential knowledge, which however must be made available by adequate processing of information. For this purpose the flood of information must yield relevant, i.e. meaningful, information, the parts of which must be correlated with one another and related to other meaningful information. A knowledge technology of this kind, which in its infancy already exists, is in the narrow sense a "semantic technology".

This already gives a glimpse of how difficult the issues are that we are dealing with here. It immediately raises the question of whether at all one can promote this term, loaded as it is from use in the humanities, to being the object of any exact science. Are the semantics that we attribute to a word or a thing not a completely subjective matter, and unamenable to objective description? Do questions of the meanings of symbols, information, knowledge and the like not precisely lie outside the reach of the modern technologies of information and communication? This consideration we already encountered in the previous chapters of this book. It is connected with one of the most important question-marks of modern science.

Only a few years ago, the question of how one might approach the semantic aspect of information by scientific and technological methods appeared completely unsolvable. However, in the meantime remarkable developments have taken place in respect of this—both theoretical and practical. As far as the theoretical side of the problem is concerned, it has become clear that there are no simple algorithms in semantics, and the search for a unified concept must be abandoned in favour of a multiple one. It is now generally recognised that the meaning content of a piece of information is never a fixed quantity, for which a unique measure can be stated, but instead is dependent upon its evaluation by the recipient and thus from that individual expectations. All these aspect have been discussed in Chap. 5.

Nevertheless, we have also seen that the criteria according to which a recipient processes some piece of information, can be classified according to general aspects. We remember, for example, that a recipient can evaluate a piece of information according to its novelty value. Alternatively, the information could be assessed according to the practical consequences that it entails for the recipient's actions. A further aspect of the recipient's assessment could be the complexity of the information, that is, the amount of its semantic content. By this procedure, the meaning that some given information has for the recipient finally crystallises out of all possible value criteria. Moreover, the recipient will give different weights to the various criteria, so that ultimately the meaning of the information takes on its highly individual and subjective character.

We thus arrive at the point where we recognise that the evaluation of information indeed has superordinate aspects that determine the essence of meaningful information, so to speak, the "meaning of meaning". These aspects constitute, as outlined in Sect. 5.3, the elements of a semantic code. This is to be understood as follows: the elements represent the pool of evaluation criteria that a recipient applies for the evaluation of the information given. However, as the elements contribute from the recipients's point of view with different weights to the meaning of the information in question, the information gets the character of uniqueness and particularity.

The concept of meaning developed here will, together with other concepts, one day belong to the tools in trade of a future technology of knowledge; this must be so if we are to handle the increasing flow of information in a constructive way. Thus in computer technology much effort is being dedicated to the development of intelligent search engines (programmes) that are able to locate information according to criteria of meaning. First steps in this direction has been the extension of the internet that is generally known as "semantic web" or "web 2.0".

For science and engineering as well, the technology of knowledge is set to become increasingly important. Science and engineering themselves are in danger of becoming submerged in the flood of scientific and technical information. The extent that this problem has already reached is illustrated by biomedical research, where (as in many other scientific disciplines) the rate of appearance of new publications has increased exponentially over the last 50 years. Already in 2005, the relevant databases showed a total of over 44 million publications registered.

Basic medical-pharmaceutical research thus faces the almost insurmountable challenge of filtering out, from a sea of specialised literature, the information that is relevant for research and, especially, for clinical practice. As the greater part of this specialised information is in verbal form, the computer-aided analysis of natural language(s), under development in computer linguistics, is of the greatest importance.

Computer linguistics belongs to the applied structural sciences. As well as the directed searching of databases, the methods of computer linguistics are being used especially for dialogue between man and machine, for example in the acoustic control of machines or for automated information-exchange on the telephone.

The automated analysis and processing of natural language, also known as "language engineering", is one of many examples that show how the traditional borders between the natural and the artificial are being eroded. In the face of this development, we shall have to take on more responsibility and not less. Even though the semantic technologies may give us unlimited access to meaningful information, it is still us who, as recipients of the information, must assess its content. We remain charged with the responsibility for deciding which information has meaning for us and which does not. The semantic technologies merely make possible the optimised bringing to bear of human knowledge; they are not in themselves creative, and they cannot replace human judgement. They can only offer us solutions that help us to achieve a goal when we address our sources of information and knowledge with the right question. This demands contextual thinking, which at the same time leaves much room for experimental thinking.

Life in the knowledge-based society will demand of us creative handling of information and knowledge. To prepare us for this, schools and universities will have to provide the necessary requisites. In the future, less will depend upon these educational institutions' providing knowledge as upon their providing techniques to give the individual the necessary competence to handle this knowledge in a constructive way.

The roof under which knowledge-related practices operate together is the technology of knowledge. It is based upon the structural sciences, in two respects. First, the structural sciences lead to knowledge of the highest generality, because its discoveries relate to the higher-order structures of reality. Secondly, the technology of knowledge, like the science of the structures of knowledge, is itself a structural science.

In knowledge-based society, therefore, the structural sciences will inevitably move into the centre of university education. As a form of science that constitutes the abstract foundation of all experience-based sciences, the structural sciences are of immense importance for the acquisition of scientific competence. One might say: thinking in the categories of the structural sciences is scientific thinking *par excellence*. Whoever masters it has genuine access to all other sciences. Only in this way can we approach the highest goal of scientific research: the computability of the world.

References

1. Black, F., Scholes, M.S.: The pricing of options and corporate liabilities. J. Political Econ. **81** (3), 637–654 (1973)
2. Cassirer, E.: Substance and Function & Einstein's Theory of Relativity. Mineola. (1980). [Original: Substanzbegriff und Funktionsbegriff, 1910]
3. Cassirer, E.: The Philosophy of Symbolic Forms, 3 vols. New Haven (1965). [Original: Philosophie der symbolischen Formen, 1923–1929]
4. Corry, L.: Writing the Ultimate Mathematical Textbook: Nicolas Bourbaki's Éléments de Mathématique. In: Robson, E., Stedall, J. (eds.): Oxford Handbook of the History of Mathematics, pp. 565–587. Oxford University Press, New York (2009)
5. Dilthey, W.: Introduction to the Human Sciences. Wayne State University Press, Detroit (1988). [Original: Einleitung in die Geisteswissenschaften, 1883]
6. Eigen, M.: Self-oganization of matter and the evolution of biological macromolecules. The Science of Nature (Naturwissenschaften) **58**, 465–523 (1971)
7. Hayek, F.A. von.: The theory of complex phenomena. In: Studies in Philosophy, Politics and Economics. London/Chicago (1967)
8. Hempel, C.G., Oppenheim, P.: Studies in the logic of explanation. Philos. Sci. **15**(2), 135–175 (1948)
9. Küppers, B.-O.: Understanding complexity. In: Beckermann, A. et al. (eds.): Emergence or Reduction?, pp. 241–256. De Gryter, Berlin/New York (1992)
10. Küppers, B.-O.: Elements of a semantic code. In: Küppers, B.-O. et al. (eds.): Evolution of Semantic Systems, pp. 67–85. Springer, Berlin/Heidelberg (2013)
11. Lévi-Strauss, C.: Structural Anthropology. Basic Books, New York (1963). [Original: Anthropologie structurale, 1958]
12. Luhmann, N.: Introduction to Systems Theory. Polity Press, Cambridge (2013). [Original: Einführung in die Systemtheorie, 2002]
13. Mandelbrot, B.: Fractals and Scaling in Finance. Springer, New York (1997)
14. Merton, R.C.: Theory of rational option pricing. Bell J. Econ. Manag. Sci. **4**, 141–183 (1973)
15. Mises, R. von: Positivism: A Study in Human Understanding. Dover, Mineola, New York (1968). [Original: Kleines Lehrbuch des Positivismus, 1939]
16. Rickert, H.: Science and History. Van Nostrand Co., New York (1962). [Original: Kulturwissenschaft und Naturwissenschaft, 1926]
17. Snow, C.P.: The Two Cultures. Cambridge University Press, Cambridge (2012)
18. Weizsäcker, C.F. von: Unity of Nature. Farrar Straus Giroux, New York (1980). [Original: Einheit der Natur, 1971]
19. Weyl, H.: Philosophy of Mathematics and Natural Science. Princeton University Press, Princton (2009). [Original: Philosophie der Mathematik und Naturwissenschaft, 1927]
20. Windelband, W.: History and natural science. Hist. Theory **19**(21), 169–185 (1980). [Original: Geschichte und Naturwissenschaft, 1926]

Figure Sources

Chapter 1

    Frontispiece: Wikipedia

Chapter 2

    Frontispiece: CERN

Chapter 3

    Frontispiece: NASA/JPL-Caltech/MSSS
    Fig. 5: Wikipedia: Madprime
    Fig. 10: iStockphoto/Thinkstock/Getty Images and Bernd-Olaf Küppers

Chapter 4

    Frontispiece: Wikipedia

Chapter 5

    Frontispiece: Comstock/Thinkstock/Getty Images

Chapter 6

    Frontispiece: Hemera//Thinkstock/Getty Images
    Fig. 12a: NASA
    Fig. 12b: Digital Vision/Thinkstock/Getty Images
    Fig. 12c: AEROSERVICE/SPL/Agentur Focus
    Fig. 12d: Wikipedia: U.S. Airforce, Foto Edward Aspera Jr
    Fig. 12e: Wikipedia: Dartmouth College
    Fig. 12f: Hemera/Thinkstock/Getty Images
    Fig. 13b: Biochemical Pathways (Poster). Boehringer Mannheim
    Fig. 14: Wikipedia: Benjah-bmm27
    Fig. 15: Richard F. Voss/IBM Research
    Fig. 16a: Dietmar Saupe, Konstanz
    Fig. 16b: Dietmar Saupe, Konstanz
    Fig. 18a: iStockphoto/Thinkstock/Getty Images
    Figures 23–28: Heinz-Otto Peitgen and Peter H. Richter, Bremen

© Springer International Publishing AG 2018   187
B.-O. Küppers, *The Computability of the World*, The Frontiers Collection,
https://doi.org/10.1007/978-3-319-67369-1

Chapter 7

Frontispiece: Digital Vision/Thinkstock/Getty Images
Fig. 29: modified after: Richard E. Dickerson and Irving Geis: Chemistry, Matter and the Universe, 1976

Chapter 8

Frontispiece: Saxon State and University Library Dresden
Fig. 37 (left): Wikipedia
Fig. 37 (right): German Museum Munich
Fig. 39a: Hemera/Thinkstock/Getty Images
Fig. 39b: Dan McCoy—Rainbow/Getty Images

Chapter 9

Frontispiece: iStockphoto/Thinkstock/Getty Images

Author Index

© Springer International Publishing AG 2018
B.-O. Küppers, *The Computability of the World*, The Frontiers Collection,
https://doi.org/10.1007/978-3-319-67369-1

Subject Index

© Springer International Publishing AG 2018

B.-O. Küppers, *The Computability of the World*, The Frontiers Collection, https://doi.org/10.1007/978-3-319-67369-1

Titles in This Series

Quantum Mechanics and Gravity
By Mendel Sachs

Quantum-Classical Correspondence
Dynamical Quantization and the Classical Limit
By A.O. Bolivar

Knowledge and the World: Challenges Beyond the Science Wars
Ed. by M. Carrier, J. Roggenhofer, G. Küppers and P. Blanchard

Quantum-Classical Analogies
By Daniela Dragoman and Mircea Dragoman

Quo Vadis Quantum Mechanics?
Ed. by Avshalom C. Elitzur, Shahar Dolev and Nancy Kolenda

Information and Its Role in Nature
By Juan G. Roederer

Extreme Events in Nature and Society
Ed. by Sergio Albeverio, Volker Jentsch and Holger Kantz

The Thermodynamic Machinery of Life
By Michal Kurzynski

Weak Links
The Universal Key to the Stability of Networks and Complex Systems
By Csermely Peter

The Emerging Physics of Consciousness
Ed. by Jack A. Tuszynski

Quantum Mechanics at the Crossroads
New Perspectives from History, Philosophy and Physics
Ed. by James Evans and Alan S. Thorndike

© Springer International Publishing AG 2018
B.-O. Küppers, *The Computability of the World*, The Frontiers Collection,
https://doi.org/10.1007/978-3-319-67369-1

Extreme States of Matter
On Earth and in the Cosmos
By Vladimir E. Fortov

Searching for Extraterrestrial Intelligence
SETI Past, Present, and Future
Ed. by H. Paul Shuch

Essential Building Blocks of Human Nature
Ed. by Ulrich J. Frey, Charlotte Störmer and Kai P. Willführ

Mindful Universe
Quantum Mechanics and the Participating Observer
By Henry P. Stapp

Principles of Evolution
From the Planck Epoch to Complex Multicellular Life
Ed. by Hildegard Meyer-Ortmanns and Stefan Thurner

The Second Law of Economics
Energy, Entropy, and the Origins of Wealth
By Reiner Kümmel

States of Consciousness
Experimental Insights into Meditation, Waking, Sleep and Dreams
Ed. by Dean Cvetkovic and Irena Cosic

Elegance and Enigma
The Quantum Interviews
Ed. by Maximilian Schlosshauer

Humans on Earth
From Origins to Possible Futures
By Filipe Duarte Santos

Evolution 2.0
Implications of Darwinism in Philosophy and the Social and Natural Sciences
Ed. by Martin Brinkworth and Friedel Weinert

Probability in Physics
Ed. by Yemima Ben-Menahem and Meir Hemmo

Chips 2020
A Guide to the Future of Nanoelectronics
Ed. by Bernd Hoefflinger

From the Web to the Grid and Beyond
Computing Paradigms Driven by High-Energy Physics
Ed. by René Brun, Frederico Carminati and Giuliana Galli-Carminati

The Language Phenomenon
Human Communication from Milliseconds to Millennia
Ed. by P.-M. Binder and K. Smith

The Dual Nature of Life
Interplay of the Individual and the Genome
By Gennadiy Zhegunov

Natural Fabrications
Science, Emergence and Consciousness
By William Seager

Ultimate Horizons
Probing the Limits of the Universe
By Helmut Satz

Physics, Nature and Society
A Guide to Order and Complexity in Our World
By Joaquín Marro

Extraterrestrial Altruism
Evolution and Ethics in the Cosmos
Ed. by Douglas A. Vakoch

The Beginning and the End
The Meaning of Life in a Cosmological Perspective
By Clément Vidal

A Brief History of String Theory
From Dual Models to M-Theory
By Dean Rickles

Singularity Hypotheses
A Scientific and Philosophical Assessment
Ed. by Amnon H. Eden, James H. Moor, Johnny H. Søraker and Eric Steinhart

Why More Is Different
Philosophical Issues in Condensed Matter Physics and Complex Systems
Ed. by Brigitte Falkenburg and Margaret Morrison

Questioning the Foundations of Physics
Which of Our Fundamental Assumptions Are Wrong?
Ed. by Anthony Aguirre, Brendan Foster and Zeeya Merali

It From Bit or Bit From It?
On Physics and Information
Ed. by Anthony Aguirre, Brendan Foster and Zeeya Merali

How Should Humanity Steer the Future?
Ed. by Anthony Aguirre, Brendan Foster and Zeeya Merali

Trick or Truth?
The Mysterious Connection Between Physics and Mathematics
Ed. by Anthony Aguirre, Brendan Foster and Zeeya Merali

The Challenge of Chance
A Multidisciplinary Approach from Science and the Humanities
Ed. by Klaas Landsman, Ellen van Wolde

Quantum [Un]Speakables II
Half a Century of Bell's Theorem
Ed. by Reinhold Bertlmann, Anton Zeilinger

Energy, Complexity and Wealth Maximization
Ed. by Robert Ayres

Ancestors, Territoriality and Gods
A Natural History of Religion
By Ina Wunn, Davina Grojnowski

Space,Time and the Limits of Human Understanding
Ed. by Shyam Wuppuluri, Giancarlo Ghirardi

Information and Interaction
Eddington, Wheeler, and the Limits of Knowledge
Ed. by Ian T. Durham, Dean Rickles

The Technological Singularity
Managing the Journey
Ed. by V. Callaghan, J. Miller, R. Yampolskiy, S. Armstrong

How Can Physics Underlie the Mind?
Top-Down Causation in the Human Context
By George Ellis

The Unknown as an Engine for Science
An Essay on the Definite and the Indefinite
Hans J. Pirner

CHIPS 2020 Vol. 2
New Vistas in Nanoelectronics
Ed. by Bernd Hoefflinger

Printed in the United States
By Bookmasters